Dynamics

for Undergraduates

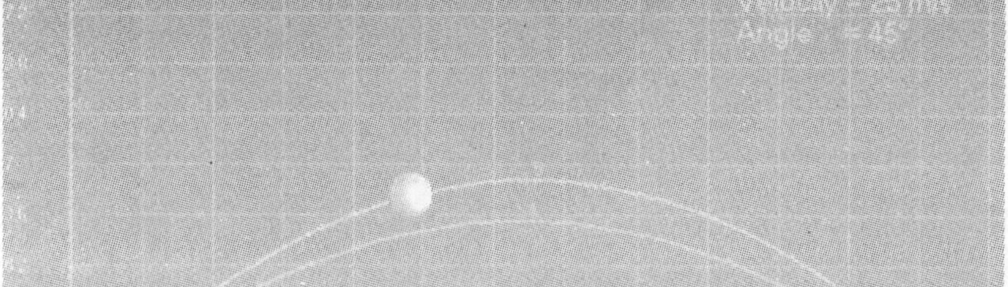
Velocity = 25 m/s
Angle = 45°

Dynamics for Undergraduates

Neeraj Pant PhD
Associate Professor
Department of Mathematics
National Defence Academy
Khadakwasla, Pune, MH
(AICTE Awarded Author)

AN Srivastava PhD
Professor and Head
Department of Mathematics
National Defence Academy
Khadakwasla, Pune, MH

CBS

CBS Publishers & Distributors Pvt Ltd
New Delhi • Bengaluru • Pune • Kochi • Chennai

Dynamics

for Undergraduates

ISBN: 978-81-239-2001-6

First Edition : 2011

Published by Satish Kumar Jain and produced by Vinod K. Jain for
CBS Publishers & Distributors Pvt Ltd
4819/XI Prahlad Street, 24 Ansari Road, Daryaganj
New Delhi 110 002, India.
Ph: 23289259, 23266861, 23266867
Fax: 011-23243014

Website: www.cbspd.com
e-mail: delhi@cbspd.com
cbspubs@vsnl.com
cbspubs@airtelmail.in.

Branches

- Bengaluru: Seema House 2975, 17th Cross, K.R. Road, Banasankari 2nd Stage, Bengaluru 560 070, Karnataka
  Ph: +91-80-26771678/79 Fax: +91-80-26771680 e-mail: bangalore@cbspd.com
- Pune: Bhuruk Prestige, Sr. No. 52/12/2+1+3/2 Narhe, Haveli (Near Katraj-Dehu Road Bypass), Pune 411 051, Maharashtra
  Ph: 020-64704058, 64704059, 32342277 Fax: +91-020-24300160 e-mail: pune@cbspd.com
- Kochi: 36/14 Kalluvilakam, Lissie Hospital Road, Kochi 682 018, Kerala
  Ph: +91-484-4059061-65 Fax: +91-484-4059065 e-mail: cochin@cbspd.com
- Chennai: 20, West Park Road, Shenoy Nagar, Chennai 600 030, Tamil Nadu
  Ph: +91-44-26260666, 26208620 Fax: +91-44-45530020 email: chennai@cbspd.com

Printed at India Binding House, Noida, UP

Preface

It gives us immense pleasure in bringing out the first edition of the book ***Dynamics*** *for Undergraduates*. The book meets the requirements of all undergraduate students of Indian universities and institutions where graduation level dynamics is the part of the curriculum. The book is written in such a manner that the students will develop keen interest in the subject.

This book encompasses the following salient features:

- Definitions, proofs of articles and theorems are given in a lucid manner with appropriate diagrams.
- Important terminology is highlighted.
- All the important formulae are kept in the box prior to illustrative examples so that the reader can recapitulate them before solving the problems.
- Algorithm of most of the multistep problems is incorporated.
- Many difficult problems are worked out with alternative methods.
- Review exercises with suitable hints are given.
- Applications of dynamics in different areas from defence to playground are given.
- Vital for competitive examinations conducted by UPSC and State PSCs.

We shall feel highly obliged for any suggestions and constructive criticism for the improvement of subsequent editions of the book.

We are thankful to our publishers, CBS Publishers & Distributors, New Delhi, and in particular to Mr YN Arjuna, Senior Director–Publishing, and his team for bringing out the much-needed book on dynamics in its present form in a short span of time.

Neeraj Pant • AN Srivastava

Contents

1 Impact

INTRODUCTION

The knowledge of impact is of paramount importance in every sphere of life right from war to play ground. The knowledge of impact theory is immensely used in understanding snooker, billiard, cricket, lawn tennis and apart from these it is also used in understanding the concept of internal ballistics of gun. Thus in light of the importance of impact theory, we shall study the theory of impact and its consequences.

§ 1. Impact

When two bodies strike or collide against each other, the state of colliding is known as their impact. Impact is of two types (i) **Direct Impact** and (ii) **Oblique Impact**.

Direct Impact

Impact between two bodies is said to be **direct** when the direction of motion of each colliding body just prior to impact is along the common normal at their point of contact (Fig. 1.1).

Note: (*i*) *Common normal is a line joining centre of the spherical bodies.*

(*ii*) *Direction of Common normal at the point of contact is known as Line of impact.*

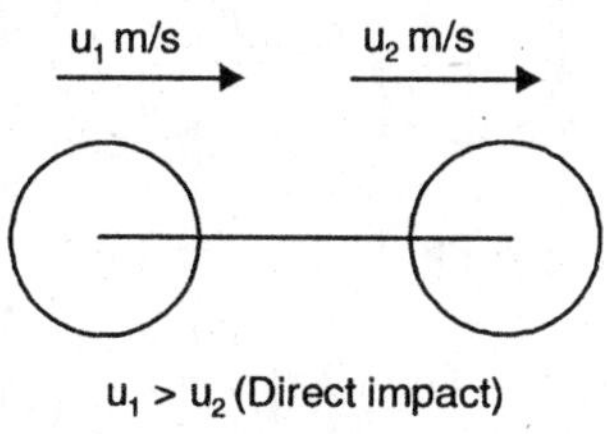

Fig. 1.1

Oblique Impact

Impact between two bodies is said to be **Oblique** when the direction of motion of one or both colliding bodies just prior to impact is not along the common normal at their point of contact (Fig 1.2).

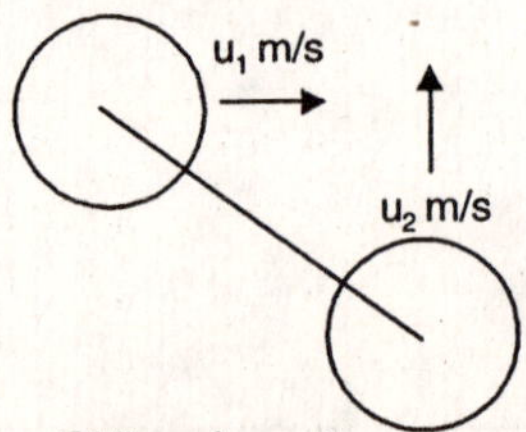

Fig. 1.2

The phenomenon of impact is completely dealt with the help of following two laws:

(i) Law of conservation of momentum (LCM) (ii) Newton's experimental law (NEL) /Newton's law of Restitution.

§ 2. Law of conservation of momentum (LCM) for direct impact. It states that the total momenta before impact is equal to the total momenta after impact, i.e.

$$m_1u_1 + m_2u_2 = m_1v_1 + m_2v_2 \tag{A}$$

where u_1, u_2 are the velocities before impact and v_1, v_2 are the velocities after impact for bodies of masses m_1 and m_2 respectively.

Proof. When two bodies A and B collide with each other, then

(i) By Newton's third Law the action (F_1) of A on B is equal and opposite to action (F_2) of B on A i.e.

$$F_1 = -F_2 \tag{i}$$

(ii) Let δt be the time of contact of A and B in the process of impact. Then impulse of action of A and B = $F_1\ \delta t = m_1(v_1 - u_1)$ (ii)

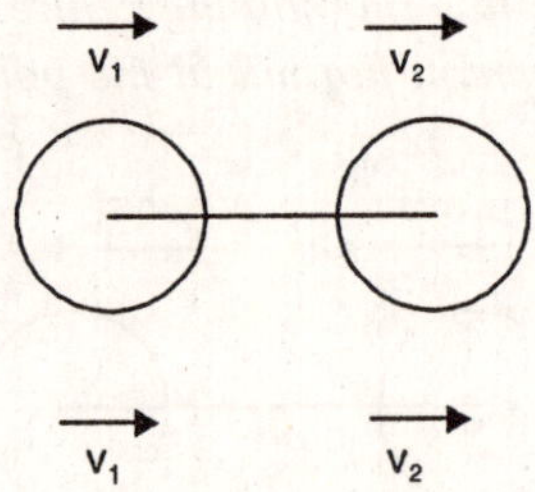

Fig. 1.3 ($u_1 > u_2$)

and impulse of action of B and A = $F_2\ \delta t = m_2(v_2 - u_2)$ (iii)

Adding (ii), (iii) and in view of (i) we get

$$m_1u_1 + m_2u_2 = m_1v_1 + m_2v_2$$

§ 3. (ii) Newton's experimental law (NEL) for direct impact. Newton observed by experiments that when two bodies collide then their relative velocity after impact is in a constant ratio of their relative velocity before impact and is in opposite direction. This constant ratio is known as **coefficient of restitution** and is denoted by ***e***.

Mathematically Newton's experimental law (**NEL**) states that

$$v_2 - v_1 = -e(u_2 - u_1) \qquad \text{(B)}$$

Coefficient of restitution (*e*) when two bodies collide each other, they remain in contact with each other for a very short duration before they separate. This short duration in which the impact lasts is divided into two parts of periods (i) **period of compression** (ii) **period of restitution. During the period of compression** the two bodies press each other and their surfaces at the contact zone get slightly **compressed.** When **compression is maximum then the period of restitution starts. During the period of** restitution the two bodies try to regain their shape (Elasticity property) and separate each other. This short duration in which the impact lasts depends on coefficient of restitution of the bodies. It is defined as:

When two bodies collide then their relative velocity after impact is in a constant ratio of their relative velocity before impact and is in opposite direction. This constant ratio is known as coefficient of restitution or modulus of restitution and is denoted by **e.**

$$e = -\left(\frac{v_2 - v_1}{u_2 - u_1}\right)$$

Note:

(*i*) *Coefficient of restitution (e) depends on the material of the colliding bodies.*

(*ii*) *It does neither depend on masses nor velocities of the bodies.*

(*iii*) *The value of e lies between* 0 *and* 1. *For perfectly elastic bodies* $e = 1$ *and for perfectly inelastic bodies* $e = 0$.

(*iv*) *Hereafter Law of conservation of momentum and Newton's experimental law may be abbreviated as LCM and NEL respectively.*

By using **NEL, LCM** and the characteristics of **coefficient of restitution (e)** we will study the direct impact in terms of (i) motion after impact i.e. determination of velocities after impact (ii) Change in momentum after impact (iii) Loss of kinetic energy after impact.

(i) **Determination of velocities after direct impact.** *Two bodies of masses* m_1 *and* m_2 *moving with velocities* u_1 *and* u_2 *respectively, impinge (collide) directly. If e be the coefficient of restitution, determine the velocities after impact.*

Let v_1 *and* v_2 be the velocities after impact for bodies of masses m_1 *and* m_2 respectively.

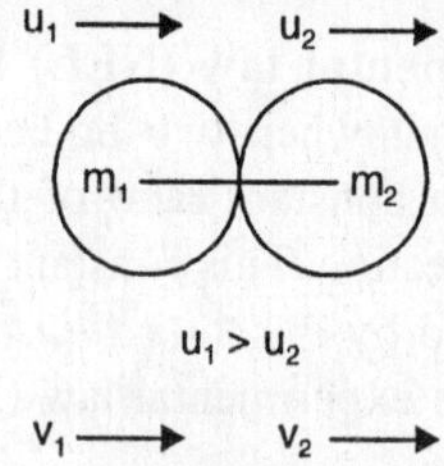

Fig. 1.4

Then by principle of **LCM**

Total momenta after impact = Total momenta before impact

$$\Rightarrow \quad m_1v_1 + m_2v_2 = m_1u_1 + m_2u_2 \qquad \text{... (i)}$$

And by principle of **NEL**

$$v_2 - v_1 = -e(u_2 - u_1) \qquad \text{... (ii)}$$

Multiplying (ii) by m_2 and then subtracting to (i) we get;

$$v_1 = \frac{m_2(1+e)u_2 + (m_1 - em_2)u_1}{m_1 + m_2} \qquad \text{... (iii)}$$

Similarly, multiplying (ii) by m_1 and then adding to (i) we get;

$$v_2 = \frac{m_1(1+e)u_1 + (m_2 - em_1)u_2}{m_1 + m_2} \qquad \text{... (iv)}$$

Results (iii) and (iv) give the velocities of the bodies after impact.

Corollary. *If equal and perfectly elastic balls move in the same direction having direct impact, they interchange their velocities. Put e = 1, $m_1 = m_2$ in (iii) and (iv) we get $v_1 = u_2$ and $v_2 = u_1$.*

(ii) **Change in momentum after impact:** *Two bodies of masses m_1 and m_2 moving with velocities u_1 and u_2 respectively, impinge (collide) directly. If e be the coefficient of restitution, show that momentum lost by one body is equal to gained by other body.*

In view of (iii) and (iv) we have,

Change in momentum of m_1 mass after impact $= m_1(v_1 - u_1)$

$$= m_1\left[\frac{m_2(1+e)u_2 + (m_1 - em_2)u_1}{m_1 + m_2} - u_1\right]$$

$$= -\frac{m_1m_2\left[(u_1 - u_2)(1+e)\right]}{m_1 + m_2} \qquad \text{... (v)}$$

Negative sign of right hand side shows that the momentum is lost by the body of mass m_1.

Similarly, change in momentum of m_2 mass after impact $= m_2(v_2 - u_2)$

$$= m_2\left[\frac{m_1(1+e)u_1 + (m_2 - em_1)u_2}{m_1 + m_2} - u_2\right]$$

$$= \frac{m_1 m_2\left[(u_1 - u_2)(1+e)\right]}{m_1 + m_2} \quad \text{... (vi)}$$

positive sign of right hand side shows that the momentum is gained by the body of mass m_2.

Thus in view of (v) and (vi) we have,

Momentum lost by the body of mass m_1 = momentum gained by the body of mass m_2

Change of momentum of the body is also known as **impulse of blow** on the body.

Corollary: *If equal and perfectly elastic balls moving in same direction having direct impact, their change of momentum is numerically equal to* $m(u_1 - u_2)$. *[Put* $e = 1$, $m_1 = m_2 = m$, *in (v) and (vi).]*

(iii) **Loss of kinetic energy due to direct impact:** *Two bodies of masses* m_1 *and* m_2 *moving with velocities* u_1 *and* u_2 *respectively, impinge directly. If e be the coefficient of restitution, determine the loss of kinetic energy due to direct impact.*

Let v_1, v_2 be the velocities after impact for bodies of masses m_1 and m_2 respectively.

Then by principle of **LCM**

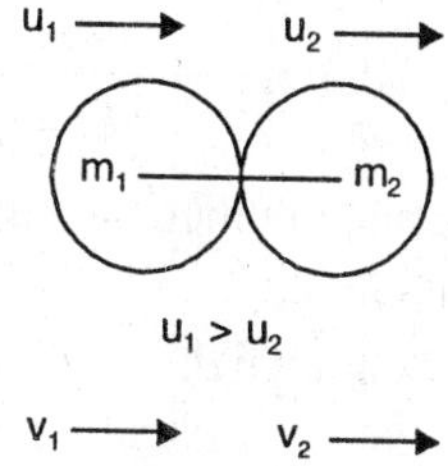

Fig. 1.5

Total momenta after impact = Total momenta before impact

$$\Rightarrow \quad m_1 v_1 + m_2 v_2 = m_1 u_1 + m_2 u_2 \quad \text{...(i)}$$

And by principle of **NEL**

$$v_2 - v_1 = -e(u_2 - u_1) \quad \text{...(ii)}$$

Adding square of (i) and square of (ii) Multiplied by m_1m_2 we get;

$$(m_1v_1+m_2v_2)^2+m_1m_2(v_2-v_1)^2=(m_1u_1+m_2u_2)^2+m_1m_2e^2(u_2-u_1)^2$$

$$\Rightarrow\ m_1(m_1+m_2)v_1^2+m_2(m_1+m_2)v_2^2$$

$$=(m_1u_1+m_2u_2)^2+m_1m_2(u_2-u_1)^2-m_1m_2(1-e^2)(u_2-u_1)^2$$

$$\Rightarrow\ m_1(m_1+m_2)v_1^2+m_2(m_1+m_2)v_2^2$$

$$=m_1(m_1+m_2)u_1^2+m_2(m_1+m_2)u_2^2-m_1m_2(1-e^2)(u_2-u_1)^2 \quad \text{... (iii)}$$

Dividing (iii) by (m_1+m_2) we get,

$$m_1v_1^2+m_2v_2^2=m_1u_1^2+m_2u_2^2-\frac{m_1m_2}{m_1+m_2}(1-e^2)(u_2-u_1)^2$$

$$\Rightarrow\ m_1v_1^2+m_2v_2^2-(m_1u_1^2+m_2u_2^2)=-\frac{m_1m_2}{m_1+m_2}(1-e^2)(u_2-u_1)^2$$

Multiplying by 1/2 we get,

$$\left(\frac{1}{2}m_1v_1^2+\frac{1}{2}m_2v_2^2\right)-\left(\frac{1}{2}m_1u_1^2+\frac{1}{2}m_2u_2^2\right)=-\frac{1}{2}\frac{m_1m_2}{m_1+m_2}(1-e^2)(u_2-u_1)^2$$

$\Rightarrow$ Total KE after impact – Total KE before impact

$$=-\frac{1}{2}\frac{m_1m_2}{m_1+m_2}(1-e^2)(u_2-u_1)^2$$

$\Rightarrow$ **Change in kinetic energy** (KE) $=-\dfrac{1}{2}\dfrac{m_1m_2}{m_1+m_2}(1-e^2)(u_2-u_1)^2$ ... (iv)

Equation (iv) shows that change in kinetic energy (KE) is negative. Hence, kinetic energy is lost due to impact.

Corollary. *If two inelastic spheres have direct impact then kinetic energy lost by the impact is equivalent to the kinetic energy of a body of mass equal to half of harmonic mean of the masses and velocity is equal to their relative velocity before impact.* [*Put* $e = 0$ *in (iv) we have, KE loss due to impact* $=\dfrac{1}{2}\dfrac{m_1m_2}{m_1+m_2}(u_2-u_1)^2=\dfrac{1}{2}MV^2$ *where* $M=\dfrac{m_1m_2}{m_1+m_2}$ = *half of harmonic mean of* m_1 and m_2. $V=u_2-u_1$ = *relative velocity before impact.*]

Important results to be used

1. LCM $\Rightarrow m_1u_1 + m_2u_2 = m_1v_1 + m_2v_2$
2. NEL $\Rightarrow v_2 - v_1 = -e(u_2 - u_1)$
3. $v_1 = \dfrac{m_2(1+e)u_2 + (m_1 - em_2)u_1}{m_1 + m_2}$, $v_2 = \dfrac{m_1(1+e)u_1 + (m_2 - em_1)u_2}{m_1 + m_2}$
4. Impulse of blow on m_1 mass $= -\dfrac{m_1m_2\left[(u_1 - u_2)(1+e)\right]}{m_1 + m_2} = -$ impulse of blow on m_2 mass
5. Change in kinetic energy (KE) $= -\dfrac{1}{2}\dfrac{m_1m_2}{2m_1 + m_2}(1-e^2)(u_2 - u_1)^2$
6. For perfectly elastic bodies $e = 1$ and for perfectly inelastic bodies $e = 0$. *Where two bodies of masses m_1 and m_2 moving with velocities u_1, u_2 and* v_1, v_2 be the velocities after impact *respectively and* **e** *be the coefficient of restitution.*

ILLUSTRATIVE EXAMPLES

Example 1 Two balls of masses *m* and 2*m* are moving directly towards each other with equal speeds. If *e* = 1/2, then show that after impact the sphere of mass 2*m* remains at rest and the speed of other remains unchanged numerically.

Solution Here $u_1 = u$, $u_2 = -u$, $m_1 = 2m$, $m_2 = m$ and $e = \dfrac{1}{2}$.

By principle of **NEL** We have,

$$v_2 - v_1 = -e(u_2 - u_1) = \frac{1(-u-u)}{2} = u$$

$\Rightarrow \quad v_2 - v_1 = u \qquad \ldots (1)$

By principle of **LCM** we have,

$$m_1u_1 + m_2u_2 = m_1v_1 + m_2v_2$$

$$2mu - mu = 2mv_1 + mv_2$$

$\Rightarrow \quad v_2 + 2v_1 = u \qquad \ldots (2)$

Eq. (2) – Eq. (1) $\Rightarrow 3v_1 = 0$

$\Rightarrow \quad v_1 = 0$ sphere of mass m comes to rest.

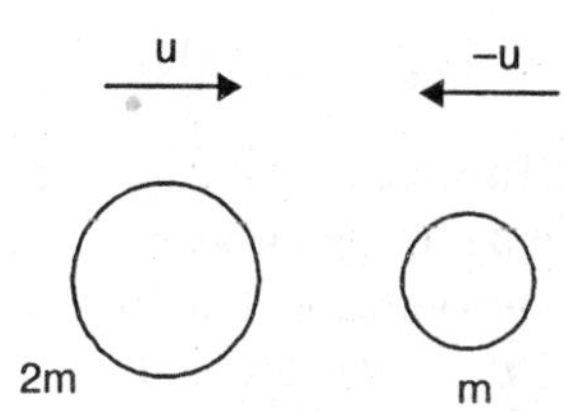

Fig. 1.6

Now putting the value of v_1 in (1) we get, $v_2 - 0 = u$

$\Rightarrow v_2 = u$ sphere of mass $2m$ returns back with same speed

Example 2 A ball impinges directly on an equal mass ball at rest; If *e* be the coefficient of restitution, show that their velocities after impact are as $(1-e):(1+e)$.

Solution Let u be the velocity of impinging ball and v_1 be its velocity after impact. Let v_2 be the velocity after impact of the ball u which was initially at rest. Let m be the mass of each ball.

Here $u_1 = u$, $u_2 = 0$ and $m_1 = m = m_2$

By principle of **NEL** we have,

$$v_2 - v_1 = -e(u_2 - u_1) = -e(0-u)$$

$$\Rightarrow v_2 - v_1 = eu \quad \ldots(1)$$

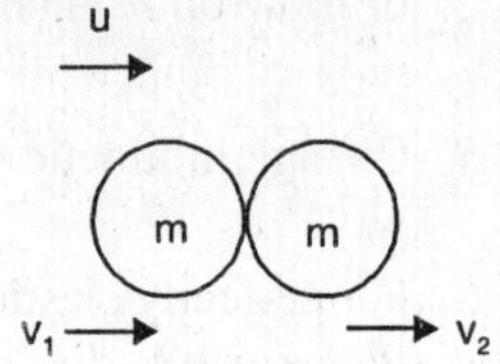

Fig. 1.7

By principle of **LCM** we have,

$$m_1u_1 + m_2u_2 = m_1v_1 + m_2v_2$$

$$mu - 0 = mv_1 + mv_2$$

$$\Rightarrow \quad v_2 + v_1 = u \quad \ldots (2)$$

Eq. (2) – Eq. (1) $\Rightarrow 2v_1 = (1-e)u$

$$\Rightarrow \quad v_1 = \frac{1}{2}(1-e)u \quad \ldots (3)$$

Now putting the value of v_1 in (1) we get,

$$v_2 - \frac{1}{2}(1-e)u = eu$$

$$\Rightarrow \quad v_2 = \frac{1}{2}(1+e)u \quad \ldots (4)$$

In view of (3) and (4) we have,

$$\therefore \quad v_1 : v_2 = (1-e):(1+e).$$

Example 3 Three spheres A, B, C having equal masses lie with their centres in a straight line. *e* = 1/ 2 between any two spheres. Sphere A is projected horizontally so as to collide directly on B, which in turn collides with C. If C moves off with the speed 27 m/s, find the projection speed of A.

Solution Let u be the velocity of projection of A and m be the mass of each sphere.

Impact between A and B

Here $u_1 = u, u_2 = 0$ and $m_1 = m = m_2$

and v_A, v_B be the velocities after impact respectively

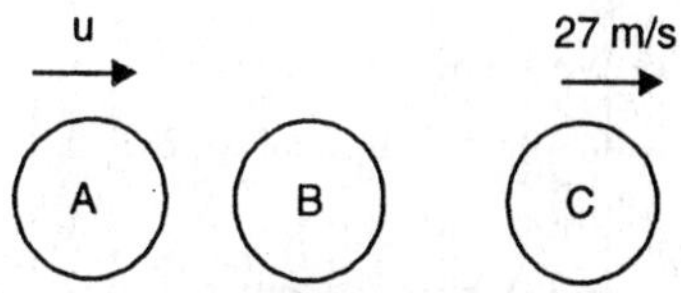

Fig. 1.8

By **principle of NEL** we have,

$$v_B - v_A = -e(0-u) = -\frac{1(-u)}{2} = \frac{1}{2}u \quad \dots (1)$$

By **principle of LCM** we have,

$$m_1u_1 + m_2u_2 = m_1v_1 + m_2v_2$$

$$mu + m \times 0 = mv_A + mv_B$$

$$\Rightarrow \quad v_B + v_A = u \quad \dots (2)$$

Eq. (2) – Eq. (1) $\Rightarrow 2v_A = \frac{u}{2}$

$\Rightarrow \quad v_A = \frac{u}{4}$ sphere A will move with speed $\frac{u}{4}$.

Now putting the value of v_A in (1) we get, $v_B - \frac{u}{4} = \frac{1}{2}u$

$\Rightarrow \quad v_B = \frac{3}{4}u$ sphere B will strike with sphere with speed $\frac{3u}{4}$.

Impact between B and C

Here $u_1 = \frac{3u}{4}, u_2 = 0$ and $m_1 = m = m_2$ and v'_B, v'_C, be the velocities after impact *respectively*

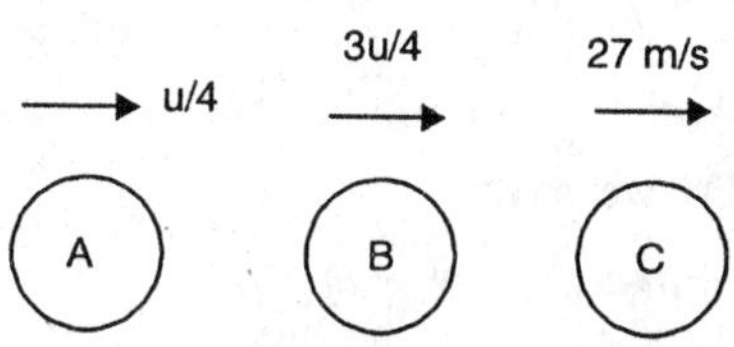

Fig. 1.9

By principle of **NEL** we have,

$$v'_C - v'_B = -e\left(0 - \frac{3u}{4}\right) = -\frac{1\left(-\frac{3u}{4}\right)}{2} = \frac{3}{8}u \qquad \dots (3)$$

By principle of **LCM** we have,

$$m_1u_1 + m_2u_2 = m_1v_1 + m_2v_2$$

$$m\frac{3u}{4} - m \times 0 = mv'_B + mv'_C$$

$$\Rightarrow \qquad v'_C + v'_B = \frac{3u}{4} \qquad \dots (4)$$

Eq. (4) – Eq. (3) $\Rightarrow 2v'_B = \frac{3u}{8}$

$\Rightarrow v'_B = \frac{3u}{16}$ sphere B will move with speed $\frac{3u}{16}$.

Now putting the value of v'_B in (3) we get, $v'_C - \frac{3u}{16} = \frac{3}{8}u$

$\Rightarrow v'_C = \frac{9}{16}u$ sphere C will move with speed $\frac{9}{16}u$

It is given that C moves with speed 27 m/s. Thus we have

$\frac{9}{16}u = 27 \Rightarrow u = 48$ m/s

Example 4 A sphere of mass *2m* moving in a straight line with speed 2*u* collides with another sphere of mass *m* moving in same direction with speed *u*. Prove that the loss of kinetic energy is $\frac{1}{3}m(1-e^2)u^2$.

Solution Here $u_1 = 2u$, $u_2 = u$, $m_1 = 2m$, $m_2 = m$.

By principle of **NEL** we have,

$$v_2 - v_1 = -e(u_2 - u_1) = -e(u - 2u) = eu$$

$$\Rightarrow \qquad v_2 - v_1 = eu \qquad \dots (1)$$

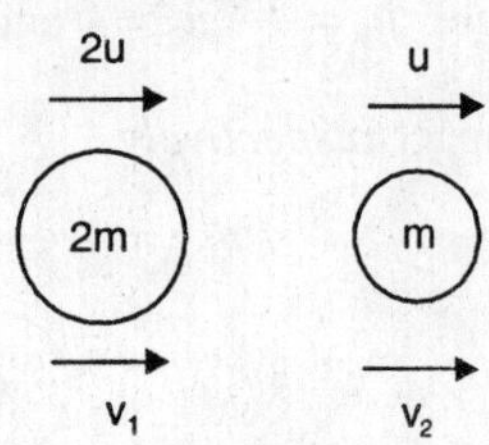

Fig. 1.10

By principle of **LCM** we have,

$$m_1u_1 + m_2u_2 = m_1v_1 + m_2v_2$$

$$2m2u + mu = 2mv_1 + mv_2$$

$$\Rightarrow \qquad v_2 + 2v_1 = 5u \qquad \dots (2)$$

Eq. (2) – Eq. (1) $\Rightarrow 3v_1 = (5-e)u$

$$\Rightarrow \qquad v_1 = \frac{(5-e)u}{3} \qquad \dots (3)$$

Now putting the value of v_1 in (1) we get,

$$v_2 - \frac{(5-e)u}{3} = eu \quad \Rightarrow$$

$$v_2 = \frac{(5+2e)u}{3} \qquad \dots (4)$$

The change in kinetic energy is

$$= \left(\frac{1}{2}2mv_1^2 + \frac{1}{2}mv_2^2\right) - \left(\frac{1}{2}2m(2u)^2 + \frac{1}{2}mu^2\right)$$

Now putting the values of v_1 and v_2 from (3) and (4) we get,

$$\text{The change in KE} = \frac{m}{2}\left[2\left(\frac{(5-e)u}{3}\right)^2 + \left(\frac{(5+2e)u}{3}\right)^2 - 8u^2 - u^2\right]$$

$$= \frac{m}{18}u^2\left[2e^2 + 50 - 20e + 4e^2 + 25 + 20e - 81\right]$$

$$= -\frac{1}{3}m(1-e^2)u^2$$

Change in kinetic energy (KE) is negative. Hence, kinetic energy is lost due to impact.

2nd Method

Change in kinetic energy (KE)

$$= -\frac{1}{2}\frac{m_1m_2}{m_1+m_2}(1-e^2)(u_2-u_1)^2$$

Putting $u_1 = 2u$, $u_2 = u$, $m_1 = 2m$, $m_2 = m$, we get,

$$= -\frac{1}{3}m(1-e^2)u^2.$$

Example 5 Three spheres A, B, C having masses *7m, 7m and m* lie with their centres in a straight line and C is between A and B. *e* = 1 between

any two spheres. Initially A and B are at rest and Sphere C is projected horizontally so as to collide directly on A. Show that it strikes A twice and B once, and the final velocities of A, B, and C are in the ratio of 21:12:1

Solution Let u be the velocity of projection of C and e = 1

Impact between C and A

Here $u_1 = u$, $u_2 = 0$, and $m_1 = m$, $m_2 = 7m$

and v_c v_A, be the velocities after impact respectively

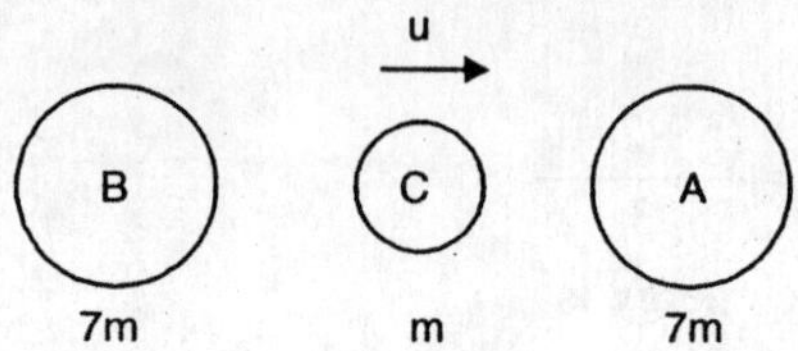

Fig. 1.11

By principle of **NEL** we have,

$$v_A - v_C = -e(0-u) = u \quad \dots (1)$$

By prineiple of **LCM** we have,

$$m_1u_1 + m_2u_2 = m_1v_1 + m_2v_2$$

$$mu + 7m \times 0 = mv_C + 7mv_A$$

$$\Rightarrow \quad 7v_A + v_C = u \quad \dots (2)$$

Eq. (2) + Eq. (1) $\Rightarrow 8v_A = 2u$

$\Rightarrow \quad v_A = \frac{u}{4}$ sphere A will move forward with speed $\frac{u}{4}$

Now putting the value of v_A in (1) we get, $\frac{u}{4} - v_C = u$

$\Rightarrow \quad v_C = \frac{-3}{4}u$ sphere C returns back with speed $\frac{3u}{4}$.

Impact between B and C

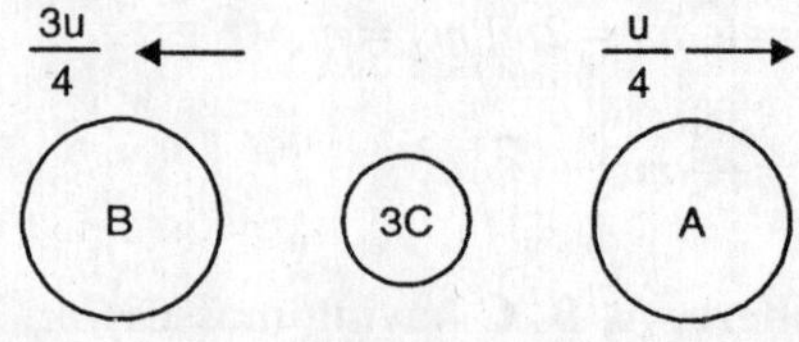

Fig. 1.12

Here $u_1 = \frac{3u}{4}$, $u_2 = 0$ and $m_1 = m$, $m_2 = 7m$

and $v'_C\ v'_B$, be the velocities after impact *respectively.*

By principle of **NEL** we have,

$$v'_B - v'_C = -e\left(0 - \frac{3u}{4}\right) = -1\left(-\frac{3u}{4}\right) = \frac{3u}{4} \qquad \ldots (3)$$

By principle of **LCM** we have,

$$m_1 u_1 + m_2 u_2 = m_1 v_1 + m_2 v_2$$

$$m\frac{3u}{4} + 7m \times 0 = mv'_C + 7mv'_B$$

$$\Rightarrow \quad 7v'_B + v'_C = \frac{3u}{4} \qquad \ldots (4)$$

Eq. (4) + Eq. (3) $\Rightarrow 8v'_B = \frac{3u}{2}$

$\Rightarrow \quad v'_B = \frac{3u}{16}$ ... (5) sphere B will move with speed $\frac{3u}{16}$

Now putting the value of v'_B in (3) we get, $\frac{3u}{16} - v'_C = \frac{3}{4}u$

$$\Rightarrow \quad v'_C = \frac{-9}{16}u \qquad \ldots (6)$$

(Negative sign means, sphere C returns back with speed $\frac{9}{16}u$).

Second Impact between C and A

Here $u_1 = \frac{9}{16}u$, $u_2 = \frac{u}{4}$ and $m_1 = m$, $m_2 = 7m$

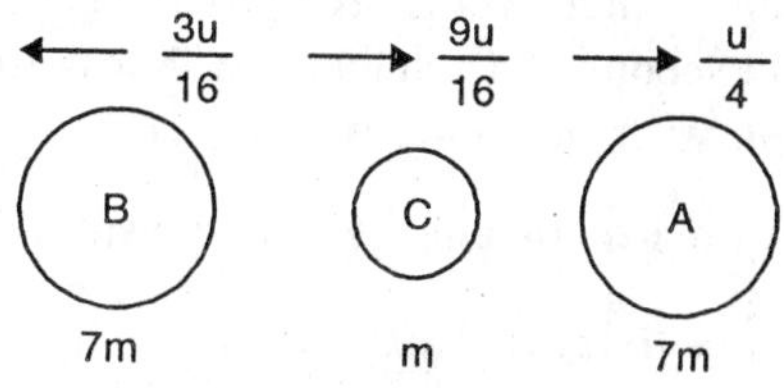

Fig. 1.13

and $v''_C\ v'_A$ be the velocities after impact *respectively*

By principle of **NEL** we have,

$$v'_A - v''_C = -1\left(\frac{u}{4} - \frac{9u}{16}\right) = \frac{5}{16}u \qquad \dots (7)$$

By principle of **LCM** we have,

$$m_1u_1 + m_2u_2 = m_1v_1 + m_2v_2$$

$$m\frac{9u}{16} + 7m\frac{u}{4} = mv''_C + 7mv'_A$$

$$\Rightarrow \quad 7v'_A + v''_C = \frac{37u}{16} \qquad \dots (8)$$

Eq. (7) + Eq. (8) $\Rightarrow 8v'_A = \dfrac{42u}{16}$

$\Rightarrow \quad v'_A = \dfrac{21u}{64}$ ... (9) sphere A will move forward with speed $\dfrac{21u}{64}$

Now putting the value of v'_A in (7) we get,

$$\frac{21u}{64} - v''_C = \frac{5}{16}u \;\Rightarrow v''_C = \frac{1}{64}u \qquad \dots (10)$$

(sphere C will move same direction as that of A with speed $\dfrac{u}{64}$)

In view of (9) and (10)

$\therefore v'_A > v''_C$ so sphere A will move faster than C and there will be no further impact between A and C.

In view of (5), (9) and (10) , the final velocities of A, B, and C are in the ratio of $v'_A : v'_B : v''_C = \dfrac{21u}{64} : \dfrac{3u}{16} : \dfrac{u}{64}$ = 21:12:1.

Example 6 A Series of *n* elastic balls whose masses are 1, e, e², e³, ... e^{n-1} at rest, with their centres on a straight line. The first is made to collide directly on the second with velocity u. Show that finally the first (n – 1) balls will move with the same velocity (1 – e) u and the last with velocity u. Prove that the kinetic energy of the system is $\frac{1}{2}\left(1-e+e^n\right)u^2$.

Solution ***Impact between first two balls***

Here $u_1 = u,\ u_2 = 0,\ m_1 = 1,\ m_2 = e.$

By principle of **NEL** we have,

$$v_2 - v_1 = -e(u_2 - u_1) = -e(0-u) = eu$$

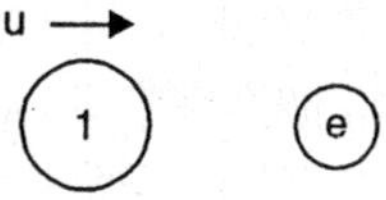

Fig. 1.14

$\Rightarrow \quad v_2 - v_1 = eu \qquad ...(1)$

By principle of **LCM** we have,

$$m_1u_1 + m_2u_2 = m_1v_1 + m_2v_2$$

$$u + e \times 0 = v_1 + ev_2$$

$\Rightarrow \quad ev_2 + v_1 = u \qquad ... (2)$

Eq. (2) + Eq. (1) $\Rightarrow (1+e)v_2 = (1+e)u$

$\Rightarrow \quad v_2 = u \qquad ... (3)$

Second ball will collide with third ball with same velocity (u).

Now putting the value of v_2 in (1) we get,

$$u - v_1 = eu \Rightarrow v_1 = (1-e)u \qquad ... (4)$$

First ball will move with the velocity $(1-e)u$.

Impact between second and third ball

In view of (3) second ball will collide with third ball with same velocity (u)

Here $u_1 = u,\ u_2 = 0,\ m_1 = e,\ m_2 = e^2$.

By principle of **NEL** we have,

$$v_4 - v_3 = -e(u_2 - u_1) = -e(0-u) = eu$$

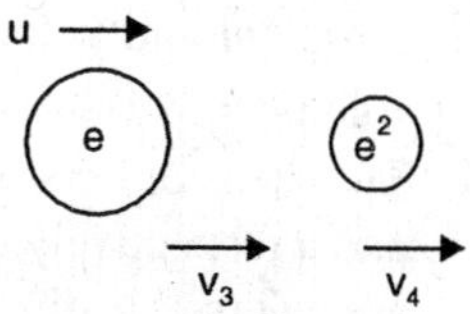

Fig. 1.15

$\Rightarrow \quad v_4 - v_3 = eu \qquad ... (5)$

By principle of **LCM** we have,

$$m_1u_1 + m_2u_2 = m_1v_1 + m_2v_2$$

$$eu + e^2 \times 0 = ev_3 + e^2v_4$$

$\Rightarrow \quad ev_3 + e^2v_4 = u \qquad ... (6)$

Solving (5) and (6) we get,

$\Rightarrow \quad v_4 = u \qquad ... (7)$

Third ball will collide with fourth ball with same velocity (u).

$$v_3 = (1-e)u \qquad ... (8)$$

Second ball will move with the velocity $(1-e)u$.

Similarly, we can proceed for further collisions.

In view of (3) and (7) the initial velocity of each collision is same and equals to initial velocity u. *Since the last ball will have no collision, therefore, it will move with velocity u.*

In view of (4) and (8) *the final velocity of each collision is same and* equal to $(1-e)u$.

$$\text{Total KE} = \frac{1}{2}[(1-e)u]^2 + \frac{1}{2}e[(1-e)u]^2 + \frac{1}{2}e^2[(1-e)u]^2$$

$$+\frac{1}{2}e^3[(1-e)u]^2 + \dots \text{ upto } (n-1) \text{ ball} + \frac{1}{2}e^{n-1}u^2$$

$$= \frac{1}{2}[(1-e)u]^2\left[1+e+e^2+e^3+\dots\right] + \frac{1}{2}e^{n-1}u^2$$

$$= \frac{1}{2}[1-e)u]^2\left[\frac{1-e^{n-1}}{1-e}\right] + \frac{1}{2}e^{n-1}u^2$$

$$= \frac{1}{2}u^2[(1-e)\left(1-e^{n-1}\right)+e^{n-1}]$$

$$= \frac{1}{2}\left(1-e+e^n\right)u^2$$

REVIEW EXERCISE 1.1

1. A sphere **A** of mass m and velocity $4u$ collides with another sphere **B** of mass $4m$ and velocity u, moving in same direction. Find e such that the sphere **A** will be at rest after the impact. **(Ans. e = 2/3)**
2. Two balls of masses m and $2m$ are moving in a straight line with speed $2u$ and u respectively, in opposite direction impinge. Prove that their velocities are reversed in direction by the impact and become $2eu$ and eu respectively, where e is the coefficient of restitution.
3. If equal and perfectly elastic balls moving in same direction have direct impact, show that they interchange their velocities after impact. (**Hint.** See cor. Determination of velocities after direct impact.)
4. Three spheres A, B, C having equal masses lie with their centres in a straight line. e = 4/5 between any two spheres. Sphere A is projected horizontally so as to collide directly on B, which in turn collides with C. If C moves of with the speed 27 m/s. Find the projection speed of A. **(Ans. u = (100/3) m/s)**
5. Three spheres A, B, C having equal masses lie with their centres in a straight line. e = 1/2 between any two spheres. Sphere A is projected horizontally so as to collide directly on B, which in turn collides with C. Find how many impacts will take place and show that ultimate speeds of the spheres are in the ratio 13:15:36. **(Ans. Three impacts)**

6. A ball of mass 8 gm moving with velocity of 10 cm/s impinges directly on another ball of mass 24 gm moving with velocity of 2 cm/s in the same direction. If e = 1/ 2, find the loss of KE.

(Ans. Loss of K. E =144 ergs)

7. If two inelastic spheres have direct impact. Show that the kinetic energy lost by the impact is equivalent to the kinetic energy of a body of mass equal to half of harmonic mean of the masses and with velocity equal to their relative velocity before impact. (**Hint.** See cor. and with loss of kinetic energy due to impact)
8. A sphere impinges directly on an equal mass sphere which is at rest. Show that the fractional loss of kinetic energy due to impact will be $\frac{1}{2}\left(1-e^2\right)$ of the original KE (**Hint.** Loss of KE/ original KE)
9. Three balls of masses m_1, m_2, m_3 are in a straight line the last two being at rest. The first ball moving in a straight line with speed u, strikes the second which in turn strikes the third. Find the velocity of the third ball after the impact. Prove that if the masses of first and third balls be given, the velocity of the third ball after impact is greatest when the masses are in geometrical, and the velocity then is equal to $\frac{m_1(1+e)^2 u}{\left(\sqrt{m_1}+\sqrt{m_3}\right)^2}$.

(**Hint.** Velocity of the third ball after impact is $v_3 = \frac{m_1(1+e)^2 u}{m_2+m_1+m_3+\frac{m_1 m_3}{m_2}}$

$m_1+m_3 =$ constant, (given) $m_2+\frac{m_1 m_3}{m_2}$, to be least for the greatest velocity and m_1, m_2, m_3, in GP if $m_2^2 = m_1 m_3$.)

§ 4. **Oblique Impact.** Impact between two bodies is said to be **Oblique** when the direction of motion of one or both colliding bodies just prior to impact is not along the common normal at their point of contact.

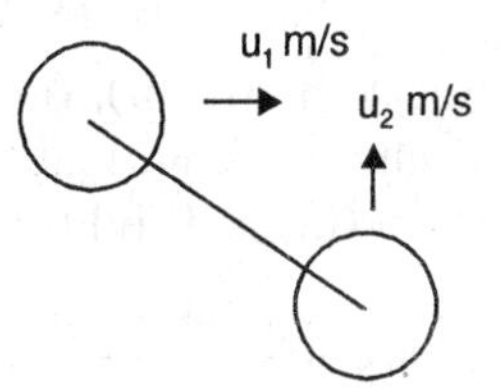

Oblique Impact

Fig. 1.16

(i) **Newton's Experimental law (NEL) on Oblique Impact.** When two bodies are in the state of oblique impact then, **(NEL) on Oblique Impact states:**

(their relative velocity resolved along their common normal before impact = −**e** (their relative velocity resolved along their common normal after impact)

(ii) **Law of conservation of momentum (LCM) on Oblique Impact.** When two bodies are in the state of oblique impact then, **(LCM) on Oblique Impact states**

(Total momenta before impact *along their common normal*) = (Total momenta after impact *along their common normal*)

§ 5. Oblique Impact of two perfectly smooth spheres. Let two perfectly smooth spheres of masses m_1 and m_2 have with velocities u_1 and u_2 respectively, impinge (collide) obliquely and make angles α and β with the line joining the centres A and B of the spheres. Let after impact v_1 and v_2 be their velocities making an angle θ and ϕ with the line AB. Since the spheres are smooth, therefore, there will be no force between them at right angle to common normal AB; on each sphere, force will act only along the line of centre i.e. along the common normal AB. Hence the components of velocity of each sphere, perpendicular to AB are unchanged due to impact, thus we have;

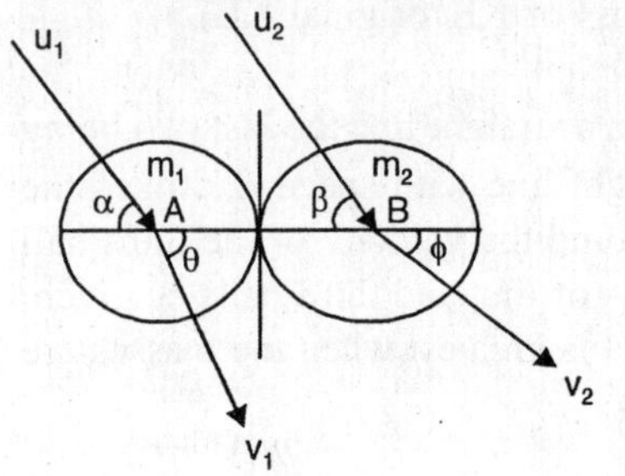

Fig. 1.17

$$v_1 \sin\theta = u_1 \sin\alpha \quad \text{... (i)}$$

$$v_2 \sin\theta = u_2 \sin\beta \quad \text{... (ii)}$$

(**Note:** Frictional force μR is along the common tangent where coefficient of friction (μ) = 0 for perfectly smooth spheres)

By NEL on Oblique Impact we have;

$$v_1 \cos\theta - v_2 \cos\theta = -e(u_1 \cos\alpha - u_2 \cos\beta) \quad \text{... (iii)}$$

By **LCM on Oblique Impact** we have;

$$m_1 v_1 \cos\theta + m_2 v_2 \cos\theta = (m_1 u_1 \cos\alpha + m_2 u_2 \cos\beta) \quad \text{... (iv)}$$

Equations (i), (ii), (iii) and (iv) are sufficient to determine the values of four unknowns, v_1, v_2, θ and ϕ.

From (iii) and (iv) [multiplying (iii) by m_2 and adding to (iv)] we get,

$$v_1 \cos\theta = \frac{m_1 u_1 \cos\alpha + m_2 u_2 \cos\beta - e m_2 (u_1 \cos\alpha - u_2 \cos\beta)}{m_1 + m_2} \quad \text{... (v)}$$

Again in view of (iii) and (iv) [multiplying (iii) by m_1 and subtracting from (iv)] we get,

$$v_2 \cos\Phi = \frac{m_1u_1\cos\alpha + m_2u_2\cos\beta + em_1(u_1\cos\alpha - u_2\cos\beta)}{m_1 + m_2} \quad \text{... (vi)}$$

On squaring and adding (i) and (v) we get the value of v_1^2 or v_1 and dividing (i) by (v) we get the value of $\tan\theta$ or θ.

On squaring and adding (ii) and (vi) we get the value of v_2^2 or v_2 and dividing (ii) by (vi) we get the value of $\tan\phi$ **or** ϕ.

Corollary 1. *If equal and perfectly elastic balls impinge obliquely, then they Interchange their velocities resolved in the direction of the line of centre.*

Put $e = 1\ m_1 = m_2$, *in (v) and (vi) we get* $(v_1\cos\theta = u_2\cos\beta)$ *and* $(v_2\cos\phi = u_1\cos\alpha)$.

Corollary 2. *If a moving smooth sphere impinges obliquely to another sphere at rest, the latter starts to move in the direction of common normal. Put* $u_2 = 0$ *in (ii), we get,* $\Phi = 0$.

Corollary 3. *If a smooth sphere moving in the direction of common normal impinges obliquely to another sphere, the former will move in the same direction i.e. along the common normal. Put* $\alpha = 0$ *in (i), we get,* $\Phi = 0$.

§ 6. Loss of kinetic energy due to oblique impact. Let two perfectly smooth spheres of masses m_1 and m_2 moving with velocities u_1 and u_2 respectively, impinge obliquely and make angles α and β with the line joining the centres A and B of the spheres. Let after impact v_1 and v_2 be their velocities making an angles θ and Φ with the line AB. Since the spheres are smooth, therefore, there will be no force between them at right angle to common normal AB.

$$v_1 \sin\theta = u_1 \sin\alpha \quad \text{... (i)}$$

$$v_2 \sin\Phi = u_2 \sin\beta \quad \text{... (ii)}$$

$$v_1\cos\theta - v_2\cos\Phi = -e(u_1\cos\alpha - u_2\cos\beta) \quad \text{... (iii)}$$

By **NEL** on Oblique Impact

$$m_1v_1\cos\theta + m_2v_2\cos\Phi = (m_1u_1\cos\alpha + m_2u_2\cos\beta) \quad \text{... (iv)}$$

By **LCM** on Oblique Impact

Square of (iv) and adding Square of (iii) Multiplied by m_1m_2 we get

$$(m_1v_1\cos\theta + m_2v_2\cos\Phi)^2 + m_1m_2(v_1\cos\theta - v_2\cos\Phi)^2$$
$$= (m_1u_1\cos\alpha + m_2u_2\cos\beta)^2 + m_1m_2e^2(u_1\cos\alpha - u_2\cos\beta)^2$$

$$\Rightarrow \quad m_1(m_1 + m_2)v_1^2\cos^2\theta + m_2(m_1 + m_2)v_2^2\cos^2\Phi$$
$$= (m_1u_1\cos\alpha + m_2u_2\cos\beta)^2 + m_1m_2(u_1\cos\alpha - u_2\cos\beta)^2$$
$$-m_1m_2(1 - e^2)(u_1\cos\alpha - u_2\cos\beta)^2$$

Introducing the term $m_1m_2\left(u_1\cos\alpha - u_2\cos\beta\right)^2$ in RHS with posivite and negative signs. (note)

$$\Rightarrow \quad m_1\left(m_1+m_2\right)v_1^2\cos^2\theta + m_2\left(m_1+m_2\right)v_2^2\cos^2\Phi$$

$$= m_1\left(m_1+m_2\right)u_1^2\cos^2\alpha + m_2\left(m_1+m_2\right)u_2^2\cos^2\beta$$

$$-m_1m_2(1-e^2)\left(u_1\cos\alpha - u_2\cos\beta\right)^2 \quad \ldots \text{(v)}$$

Dividing (v) by (m_1+m_2) we get,

$$m_1v_1^2\cos^2\theta + m_2v_2^2\cos^2\Phi$$

$$= m_1u_1^2\cos^2\alpha + m_2u_2^2\cos^2\beta - \frac{m_1m_2(1-e^2)\left(u_1\cos\alpha - u_2\cos\beta\right)^2}{m_1+m_2} \quad \ldots \text{(vi)}$$

Also squaring and adding (i) m_1 and (ii) m_2 we get,

$$m_1v_1^2\sin^2\theta + m_2v_2^2\sin^2\Phi = m_1u_1^2\sin^2\alpha + m_2u_2^2\sin^2\beta \quad \ldots \text{(vii)}$$

Now adding (vi) and (vii) we get,

$$\Rightarrow \quad m_1v_1^2 + m_2v_2^2 - (m_1u_1^2 + m_2u_2^2) = -\frac{m_1m_2(1-e^2)\left(u_1\cos\alpha - u_2\cos\beta\right)^2}{m_1+m_2}$$

Multiplying by 1/2 we get,

$$\left(\frac{1}{2}m_1v_1^2 + \frac{1}{2}m_2v_2^2\right) - \left(\frac{1}{2}m_1u_1^2 + \frac{1}{2}m_2u_2^2\right)$$

$$= -\frac{1}{2}\frac{m_1m_2(1-e^2)\left(u_1\cos\alpha - u_2\cos\beta\right)^2}{m_1+m_2}$$

$\Rightarrow$ Total KE after impact – Total KE before impact

$$= -\frac{1}{2}\frac{m_1m_2(1-e^2)\left(u_1\cos\alpha - u_2\cos\beta\right)^2}{m_1+m_2}$$

$\Rightarrow$ Change in kinetic energy (KE)

$$= -\frac{1}{2}\frac{m_1m_2(1-e^2)\left(u_1\cos\alpha - u_2\cos\beta\right)^2}{m_1+m_2} \quad \ldots \text{(viii)}$$

Equation (iv) shows that Change in kinetic energy (KE) is negative. Hence, kinetic energy is lost due to impact.

Corollary. *If two inelastic spheres have oblique impact then kinetic energy lost by the impact is equivalent to the kinetic energy of a body of mass equal* *to half of harmonic mean of the masses and with velocity equal to their relative velocity before impact in the direction of the common normal. Put e = 0 in (viii) we have, KE loss due to impact*

$$=\frac{1}{2}\frac{m_1m_2(1-e^2)\left(u_1\cos\alpha-u_2\cos\beta\right)^2}{m_1+m_2}=\frac{1}{2}MV^2$$

where $M=\dfrac{m_1m_2}{m_1+m_2}$ = half of harmonic mean of m_1 and m_2.

$V=\left(u_1\cos\alpha-u_2\cos\beta\right)=$ relative velocity before impact along common normal.

Important results to be used

1. $v_1\sin\theta=u_1\sin\alpha,\ v_2\sin\theta=u_2\sin\beta$ (Unaltered velocity components Perp. to the com. normal of each ball.)
2. (i) If $u_2=0\Rightarrow\Phi=0\Rightarrow$ If a moving smooth sphere impinges obliquely to another sphere at rest, the latter starts to move in the direction of common normal.

   (ii) If $\alpha=0\Rightarrow\theta=0\Rightarrow$ If a smooth sphere moving in the direction of common normal impinges obliquely to another sphere, the former will move in the same direction.
3. $v_1\cos\theta-v_2\cos\Phi=-e(u_1\cos\alpha-u_2\cos\beta)$ (NEL on Oblique Impact)
4. $m_1v_1\cos\theta+m_2v_2\cos\theta=(m_1u_1\cos\alpha+m_2u_2\cos\beta)$ (LCM on Oblique Impact)
5. $v_1\cos\theta=\dfrac{m_1u_1\cos\alpha+m_2u_2\cos\beta-em_2(u_1\cos\alpha-u_2\cos\beta)}{m_1+m_2}$
6. $v_2\cos\Phi=\dfrac{m_1u_1\cos\alpha+m_2u_2\cos\beta+em_1(u_1\cos\alpha-u_2\cos\beta)}{m_1+m_2}$
7. Change in kinetic energy (KE)

$$=-\frac{1}{2}\frac{m_1m_2(1-e^2)\left(u_1\cos\alpha-u_2\cos\beta\right)^2}{m_1+m_2}$$

ILLUSTRATIVE EXAMPLES

Example 1 Two perfectly elastic smooth equal balls impinge each other obliquely. Show that they interchange their velocities in the direction of the line of centres.

Solution Let two perfectly smooth balls each of mass m moving with velocities u_1 and u_2 respectively, impinge (collide) obliquely and making angles α and β with the line joining the centres A and B of the balls. let after impact v_1 and v_2 be their velocities making angles θ and Φ with the line AB. Since the balls are smooth, therefore, there will be no force between them at right angle to common normal AB; **on each ball, force will work along the line of centre** u_1 u_2 i.e. along the common normal AB. Hence the components of velocity of each ball, perpendicular to AB are unchanged due to **impact, thus we have;**

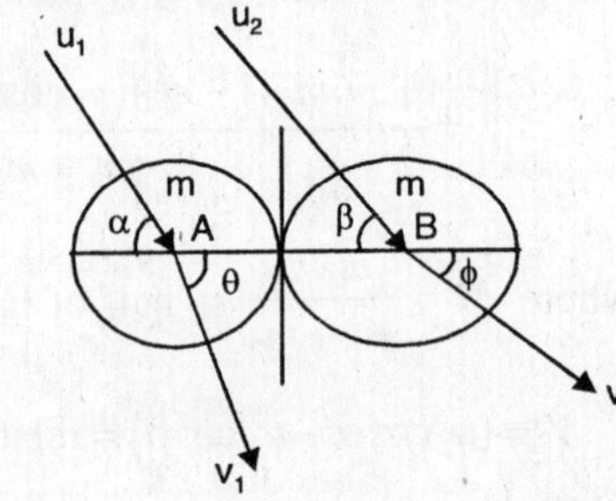

Fig. 1.18

$$v_1 \sin\theta = u_1 \sin\alpha \quad \text{... (1)}$$

$$v_2 \sin\Phi = u_2 \sin\beta \quad \text{... (2)}$$

By **NEL on Oblique Impact** we have;

$$v_1 \cos\theta - v_2 \cos\Phi = -e(u_1 \cos\alpha - u_2 \cos\beta)$$

$$\Rightarrow \quad v_1 \cos\theta - v_2 \cos\Phi = -(u_1 \cos\alpha - u_2 \cos\beta) \quad \text{...(3)}$$

(for perfectly elastic collision $e = 1$)

By **LCM on Oblique Impact** we have;

$$m_1 v_1 \cos\theta + m_2 v_2 \cos\Phi = (m_1 u_1 \cos\alpha + m_2 u_2 \cos\beta)$$

$$m v_1 \cos\theta + m v_2 \cos\Phi = (m u_1 \cos\alpha + m u_2 \cos\beta)$$

(As balls are equal $m_1 = m_2$)

$$\Rightarrow \quad v_1 \cos\theta + v_2 \cos\Phi = (u_1 \cos\alpha + u_2 \cos\beta) \quad \text{...(4)}$$

Adding (3) and (4), subtracting (3) and (4) and on simplification we get,

$$v_1 \cos\theta = u_2 \cos\beta, \; v_2 \cos\Phi = u_1 \cos\alpha$$

i.e. in the direction of the line of centres velocity after impact of first ball = velocity before impact of second ball in the same direction and in the direction of the line of centres.

Velocity after impact of second ball = velocity before impact of first ball in the same direction. Hence proved.

Example 2 A ball of mass *M* collides another ball of mass *m*, at rest. If after impact they separate in mutually perpendicular directions, prove that the coefficient of restitution is equal to the ratio of their masses.

Solution Let u_1 be the velocity before impact of the ball of mass M *making an* angle α with the line joining the centres. The ball of mass m before impact was at rest.

$\Rightarrow \quad u_2 = 0, \quad \beta = 0, \quad \dots (1)$

$v_1 \sin\theta = u_1 \sin\alpha \quad \dots (2)$

$v_2 \sin\Phi = u_2 \sin\beta \quad \dots (3)$

In view of (1) and (3) we get,

$v_2 \sin\Phi = 0 \Rightarrow \Phi = 0 \quad \dots (4)$

Thus the second ball after impact will move along common normal.

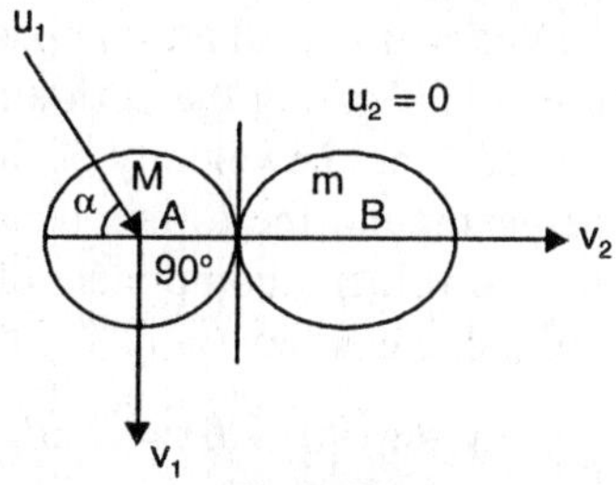

Fig. 1.19

As the first ball after impact has to move perpendicular to second ball i.e.

$\Rightarrow \quad \theta = 90°$ ... (5)

By **NEL on Oblique Impact** we have;

$$v_1 \cos 90^0 - v_2 \cos 0 = -e(u_1 \cos\alpha - 0 \times \cos 0)$$

$\Rightarrow \quad v_2 = eu_1 \cos\alpha$ ... (6)

By **LCM on Oblique Impact** we have;

$$Mv_1 \cos 90^0 + mv_2 \cos 0 = (Mu_1 \cos\alpha + m \times 0 \times \cos 0)$$

$$mv_2 = Mu_1 \cos\alpha \quad \dots (7)$$

Dividing (7) by (6) we have; $m = \frac{M}{e}$

or $e = \frac{M}{m}$ = ratio of masses of the balls.

Example 3 A ball of mass m moving with velocity u collides obliquely to another ball of mass M at rest. The original line of motion of first ball makes an angle θ with the line of the centres, at the moment of impact. Show that the ball of mass m will be deflected through a right angle if $\tan^2\theta = \frac{eM - m}{M + m}$.

Solution Let u be the velocity before impact of the ball of mass m making an angle θ with the line joining the centres. The ball of mass M before impact was at rest. Let v_2 be the velocity of Second ball of mass M, after impact.

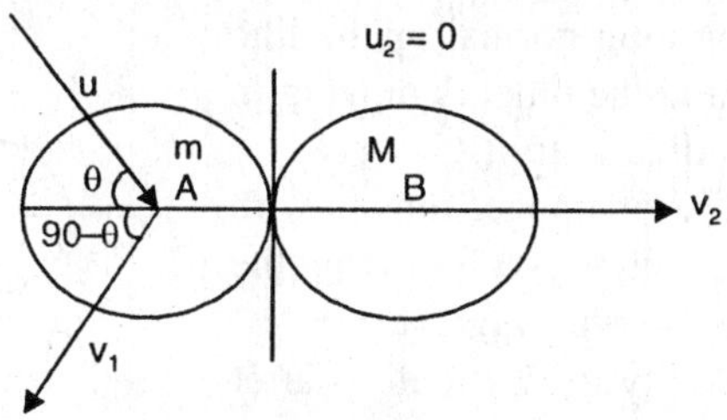

Fig. 1.20

The second ball after impact will move along common normal because the force on it during the period of impact v_2 is only along the common normal.

Let v_1 be the velocity of first ball v_1 of mass m after impact, which moves perpendicular the former direction.

The velocity component of first ball of mass m perpendicular to common normal, we have

$$v_1 \sin(90^0 - \theta) = u \sin\theta$$

$$\Rightarrow \quad v_1 \cos\theta = u \sin\theta \quad \ldots (1)$$

By **NEL on Oblique Impact** we have;

$$-v_1 \cos(90^0 - \theta) - v_2 \cos 0 = -e(u \cos\theta - 0 \times \cos 0)$$

$$\Rightarrow \quad -v_1 \sin\theta - v_2 = -eu\cos\theta \quad \ldots (2)$$

By **LCM on Oblique Impact** we have;

$$-mv_1 \cos(90^0 - \theta) + Mv_2 \cos 0 = (mu\cos\theta + M \times 0 \times \cos 0)$$

$$\Rightarrow \quad -mv_1 \sin\theta + Mv_2 = mu\cos\theta \quad \ldots (3)$$

Multiplying (2) by M and adding with (3) we get;

$$-(M+m)v_1 \sin\theta = (m - eM)u\cos\theta \quad \ldots (4)$$

Dividing (4) by (1) we get;

$$-(M+m)\tan\theta = (m - eM)\cot\theta$$

$$\Rightarrow \quad \tan^2\theta = \frac{eM - m}{M + m}.$$

Example 4 A ball impinges to another equal ball moving with same speed in a direction perpendicular to its own, the line of the centres of the balls at the instant of impact being perpendicular to the direction of motion of the second ball. If e be the coefficient of restitution, show that the direction of motion of second ball is turned through an angle $\tan^{-1}\left\{\frac{(1+e)}{2}\right\}$.

Solution Let u be the velocity before impact of each ball of mass m. One ball is moving along common normal *or* the line joining the centres and the other is moving in the direction perpendicular to it.

The second ball after impact moves with velocity v_2 making an angle $(90° - \alpha)$ with the common normal (as shown in figure).

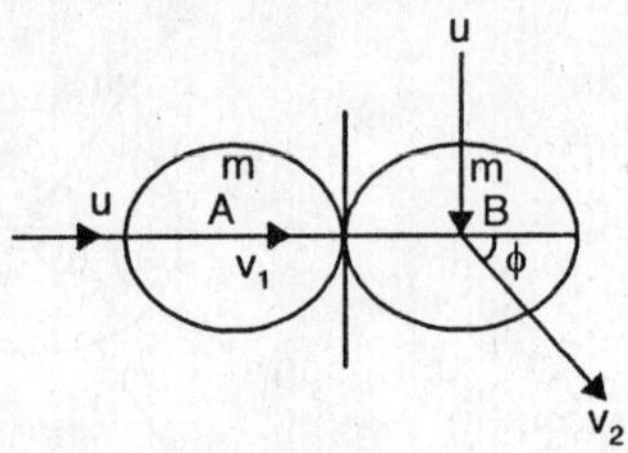

Fig. 1.21

Let v_1 be the velocity of first ball v_2 after impact, which will move along the former direction as the only force acting on it during the period of impact is also in this sense (Art 5. Corollary 3).

$\because$ **the velocity component of second ball of mass m is perpendicular to common normal,** we have;

$$v_2 \sin\Phi = u \qquad \ldots (1)$$

By NEL on Oblique Impact we have;

$$v_2 \cos\Phi - v_1 \cos 0 = -e(u\cos 90^0 - u \times \cos 0)$$

$$\Rightarrow \quad v_2 \cos\Phi - v_1 = eu \qquad \ldots (2)$$

By **LCM on Oblique Impact** we have;

$$mv_2 \cos\Phi + mv_1 \cos 0 = (mu\cos 90^0 + mu \times \cos 0)$$

$$\Rightarrow \quad v_2 \cos\Phi + v_1 = u \qquad \ldots (3)$$

Adding (2) and (3) we get;

$$v_2 \cos\Phi = \frac{1}{2}(1+e) \qquad \ldots (4)$$

Dividing (1) by (4) we get;

$$\tan\Phi = \frac{2}{(1+e)}$$

$$\text{Required angle} = 90° - \Phi = \tan^{-1}\tan\left(90^0 - \Phi\right)$$

$$= \tan^{-1}\cot\Phi = \tan^{-1}\left\{\frac{(1+e)}{2}\right\}.$$

Example 5 Two equal spheres, of radius r, move with the same speed in opposite directions in parallel lines which are at a distance d apart; prove that the direction of each deviates on impact, through right angle if $d^2(1+e) = 4er^2$, where e is the co-efficient of restitution.

Solution Let two perfectly smooth spheres each of mass m moving with velocity u impinge obliquely. Each sphere is making angle α with the line

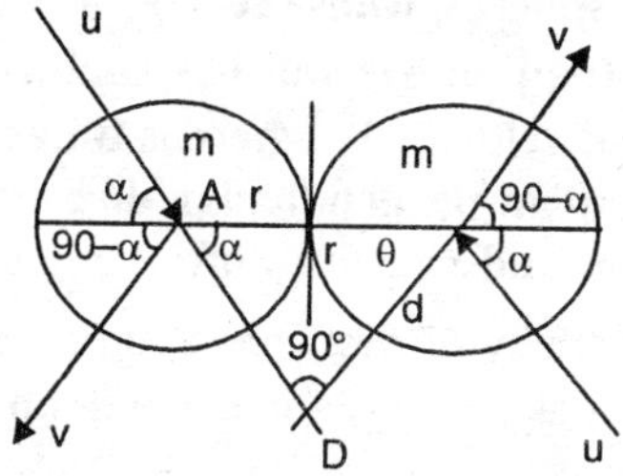

Fig. 1.22

joining the centres A and B of the spheres but there velocities are given to be in opposite directions.

Due to symmetry their velocities after impact v **will also equal in magnitude.** Let after impact each sphere has velocity v. It is given that after impact, they deviate through a right, hence after impact their directions 90° will make an angle $(90° - \alpha)$ with the v common normal AB.

$\because$ Components of velocity of each ball, perpendicular to AB are unchanged due to impact, we have

$$u \sin \alpha = v \sin (90° - \alpha)$$

$$\Rightarrow \quad u \sin \alpha = v \cos \alpha \qquad \text{... (1)}$$

By **NEL on Oblique Impact** we have;

$$-v \cos (90° - \alpha) - v \cos (90° - \alpha) = -e\{u \cos \alpha - (-u \cos \alpha)\}$$

$$\Rightarrow \quad -v \sin \alpha - v \sin \alpha = -2eu \cos \alpha$$

$$\Rightarrow \quad v \sin \alpha = eu \cos \alpha$$

$$\Rightarrow \quad v = eu \cot \alpha \qquad \text{... (2)}$$

Putting the value of v in (1) we get;

$$u \sin \alpha = eu \cot \alpha \cos \alpha$$

$$\Rightarrow \quad \tan^2 \alpha = e \qquad \text{... (4)}$$

From triangle ABD we have,

$$\sin \alpha = \frac{d}{2r}$$

$$\Rightarrow \quad \tan \alpha = \frac{d}{\sqrt{(4r^2 - d^2)}} \qquad \text{... (5)}$$

Putting the value of $\tan\alpha$ in (4) we get; $e = \dfrac{d^2}{(4r^2 - d^2)}$

$$\Rightarrow \quad d^2 (1 + e) = 4er^2.$$

Example 6 Two smooth spheres, whose centres are A and B are moving on horizontal table along the line AB and CA such that $\angle ABC$ =120⁰. The sphere A has mass m and velocity $3u$ whereas B has mass $5m$ and velocity $2u$. The spheres move off in perpendicular direction after impact. Find co-efficient of restitution and velocities after impact.

Solution Let v_1 be the velocity of first sphere after impact, which will move along the former direction as the only force acting on it during the period of impact is also in this sense.

The second sphere after impact moves with velocity v_2 making an angle 90^0 with the common normal (As in figure).

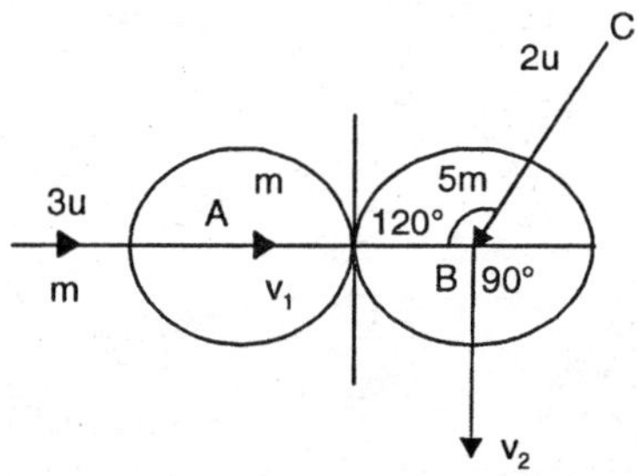

Fig. 1.23

∵ The velocity component of second ball of mass m is perpendicular to common normal, we have;

$$v_2 \sin 90^0 = 2u \sin 60^0$$

$$\Rightarrow \qquad v_2 = u\sqrt{3} \qquad \ldots (1)$$

By **NEL on Oblique Impact** we have;

$$v_2 \cos 90^0 - v_1 \cos 0 = -e(2u \cos 120^0 - 3u \times \cos 0)$$

$$\Rightarrow \qquad v_1 = 4eu \qquad \ldots (2)$$

By **LCM on Oblique Impact** we have;

$$(5mv_2 \cos 90^0 + mv_1 \cos 0) = (5m \times 2u \cos 120^0 + m \times 3u \times \cos 0)$$

$$\Rightarrow \qquad v_1 = 2u \qquad \ldots (3)$$

From (2) and (3) we get; $4eu = 2u$

$$\Rightarrow \qquad e = \frac{1}{2} \qquad \ldots (4)$$

$$v_1 = 2u,\ v_2 = u\sqrt{3} \text{ and } e = \frac{1}{2}$$

Example 7 Two equal smooth spheres of mass m are in contact on a smooth horizontal table. Third equal radius sphere of mass m_1 impinges symmetrically on them and is reduced to rest. Prove that $e = \frac{2m_1}{3m}$.

Solution The masses of the spheres with centres B and C are m each and the mass of striking sphere is m_1 and centre A. Let u be the velocity before impact of sphere with centre A in the direction of common tangent of the spheres with centres B and C, lying in contact and at rest (shown in figure).

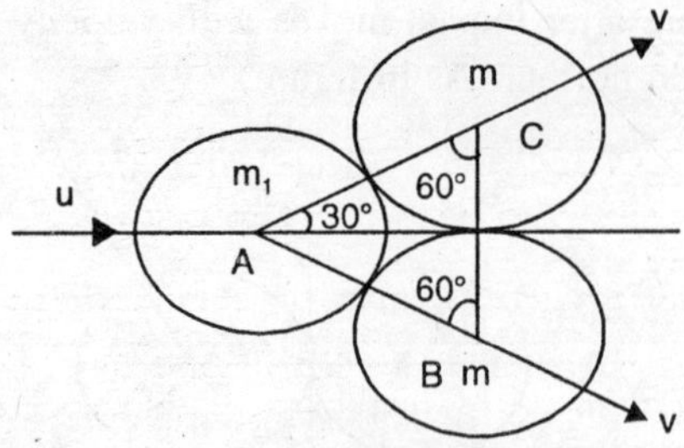

Fig. 1.24

During the period of impact there is no force on the sphere A at right angle to its original direction of motion therefore, it continues to move in original direction after impact. But we are given that A sphere at rest after impact. ΔABC is an equilateral triangle as all spheres are of equal radii and touching each other.

By **NEL on Oblique Impact** on one of the spheres and the impinging sphere

$$v - 0 = -e(0 - u\cos 30°)$$

$$\Rightarrow \quad v = eu\frac{\sqrt{3}}{2} \qquad \dots (1)$$

By **LCM on Oblique Impact** we have;

$$2mv\cos 30^0 + m_1 \times 0 = m_1 u$$

$$\Rightarrow \quad mv\sqrt{3} = m_1 u \qquad \dots (2)$$

From (1) and (2) we get; $meu\frac{\sqrt{3}}{2}\sqrt{3} = m_1 u$

$$\Rightarrow \quad e = \frac{2m_1}{3m}.$$

REVIEW EXERCISE 1.2

1. A ball of mass m moving with velocity u collides obliquely to another ball of mass nm at rest. The original line of motion of first ball makes an angle θ with the line of the centres, at the moment of impact. Show that the ball of mass m will be deflected through a right angle if $\tan^2\theta = \frac{ne-1}{n+1}$.
2. Two equal spheres of coefficient of restitution e impinge, having before impact velocities u_1, v_1 in the direction of common normal, and u_2, v_2

perpendicular to common normal. If after impact the spheres move perpendicular to each other, prove that $(u_1 + v_1)^2 + 4u_1v_1 = e^2(u_1 - v_1)^2$.

3. Two spheres of coefficient of restitution e moving in parallel directions with equal momenta impinge, prove that if the direction of motion be opposite, they will move after impact in parallel directions with equal momenta.
4. Two equal smooth spheres of mass m and radius a are in contact on a smooth horizontal table. Third sphere of mass m_l and radius c is moving in the direction of their common tangent; if e be the coefficient of restitution between them, prove that the impinging sphere will be reduced to rest if $2e = \dfrac{c^2(a+c)^2}{a^3(2a+c)}$.
5. Two equal spheres, of radius r, moves with the same speed in opposite directions in parallel lines which are at a distance c apart; prove that the direction of each deviates on impact, through right angle if $c^2 = 4er^2/(1+e)$, where e is the co-efficient of restitution.
6. Ball A moving with velocity u collides obliquely to another ball B at rest. The original line motion of ball A makes an angle θ with the line of the centres, at the moment of impact. Show that the ball A will be deflected through a right angle if their masses are in the ratio $(e\cos^2\theta - \sin^2\theta) : 1$, where e is the co-efficient of restitution.
7. If equal and perfectly elastic balls impinge obliquely. Then show that they interchange their velocities resolved in the direction of the line of centre.
8. If a moving smooth sphere impinges obliquely to another sphere at rest. Show that after impact the latter starts to move in the direction of common normal.
9. If two inelastic spheres have oblique impact. Then show that kinetic energy lost by the impact is equivalent to the kinetic energy of body of mass equal to half of harmonic mean of the masses and velocity is equal to their relative velocity before impact in the direction of the common normal.

§ 7. Direct impact of a smooth sphere on a fixed smooth surface. Let the sphere of mass m strike perpendicularly on the fixed plane at A and let its velocities be u and v respectively before and after impact (shown in Fig.).

The sphere and the plane being smooth, there is no force acting on the sphere in the direction perpendicular to the to the common normal at A. *The fixed surface may be considered to be another Body whose velocity before and after impact is zero.*

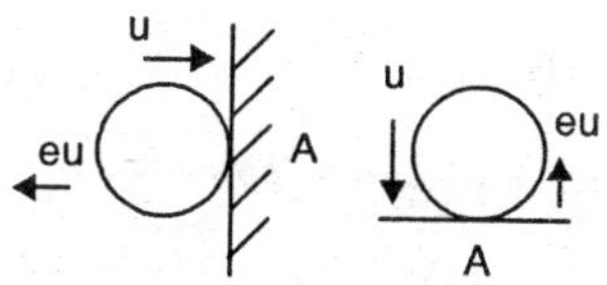

Fig. 1.25

By principle of **NEL**

$$v - 0 = -e(u - 0)$$

$\Rightarrow \quad v = -eu$

Thus after direct impact against fixed smooth surface, the sphere will move back with a velocity = ***e*** times the velocity before impact.

Impulse of blow on the surface = change of Momentum of sphere

$$= m(u - v) = mu(1 + e) \cdot$$

Loss of kinetic energy of the sphere due to impact

$$= \frac{1}{2}m(u^2 - v^2) = -\frac{1}{2}mu^2(e^2 - 1).$$

ILLUSTRATIVE EXAMPLES

Example 1 A ball falls from a height *h* above a horizontal plane, and rebounds continually and finally comes to rest. Show that the whole distance described by the ball is $\left(\frac{1+e^2}{1-e^2}\right)h$ and total time taken by ball to come at rest is $\sqrt{\left(\frac{2h}{g}\right)}\left(\frac{1+e}{1-e}\right)$ *e*, being the co-efficient of restitution.

Solution Let v be the velocity of the ball when it strike the horizontal plane first time with free fall from height h. Then by law of motion,

$v^2 = u^2 + 2gh = 2gh \quad \because u = 0$. For freefall

$\Rightarrow \quad v = \sqrt{2gh}$ ... (1)

Since the velocity after rebound = ***e*** times the velocity before impact.

$$v_1 = ev = e\sqrt{2gh} \quad \text{... (2)}$$

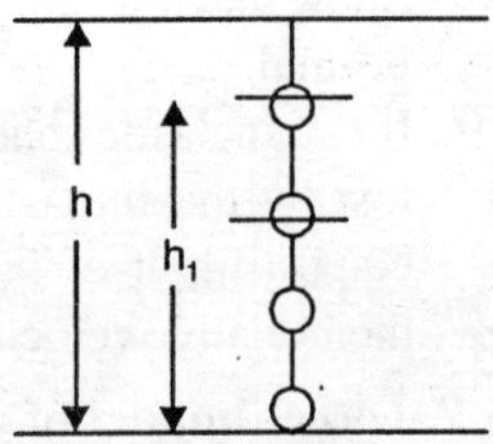

Fig. 1.26

Now v_1 acts as an initial velocity for upward motion after first bounce. Let h_1 be the height attained by the ball after first impact, we have

$0 = v_1^2 - 2gh_1$

$\Rightarrow \quad v_1^2 = 2gh_1$ ... (3)

In view of (2) and (3) we get,

$\Rightarrow \quad e^2 2gh = 2gh_1 \Rightarrow h_1 = e^2 h$ ... (4)

Similarly, $h_2 = e^2 h_1 = e^4 h$, $h_3 = e^2 h_2 = e^6 h$, and so on.

Here $h_2, h_3, \ldots$ are the heights attained by the ball after second, third ...bounces respectively.

Distance travelled between first and second impact = 2 × height attained after first bounce = $2h_1$. Similarly, Distance travelled between second and third impact = $2h_2$, and so on.

Let the total distance described by the ball be S. Therefore,

$$S = h + 2h_1 + 2h_2 + 2h_3 + \ldots + \text{ad. upto inf.}$$

$$= h(1 + 2e^2 + 2e^4 + 2e^6 + \ldots \text{ad. upto inf.})$$

$$= h\{1 + 2(e^2 + e^4 + e^6 + \ldots \text{ ad. upto inf.})\}$$

$$= h\left(1 + 2\frac{e^2}{1-e^2}\right) = \left(\frac{1+e^2}{1-e^2}\right)h,$$

(**Note.** *Sum of infinite Geometrical series* $S_\infty = \frac{a}{1-r}$).

2nd Part. Let t be the time taken by the ball when it strike the horizontal plane first time with free fall from height h. Then by law of motion,

$h = ut + \frac{1}{2}gt^2 \Rightarrow h = \frac{1}{2}gt^2$, since $u = 0$, For freefall.

$$\Rightarrow \qquad t = \sqrt{\frac{2h}{g}} \qquad \ldots (5)$$

$$v = u + gt \Rightarrow v = gt \qquad \ldots (6)$$

Let t_1 be the time taken by the ball to reach the highest point after first impact, we have $0 = ev - gt_1 \Rightarrow egt = gt_1$ using (6)

$$\Rightarrow \qquad t_1 = et \qquad \ldots (7)$$

Similarly, $t_2 = et_1 = e^2t$, $t_3 = et_2 = e^3t$, and so on.

Here $t_2, t_3, \ldots$are the time taken by the ball to reach the highest points by the ball after second, third ... bounces respectively.

Time between first and second impact = 2 × time after first impact to reach the highest point = $2t_1$. Similarly, Time between second impact and third impact = $2t_2$, and so on.

Let the total time taken by ball to come at rest be T. Therefore,

$$T = t + 2t_1 + 2t_2 + 2t_3 + \ldots + \text{ad. upto inf..}$$

$$= t(1 + 2e + 2e^2 + 2e^3 + \ldots \text{ad. upto inf.})$$

$$= t\{1 + 2(e + e^2 + e^3 + \ldots \text{ad. upto inf.})\}$$

Putting the value of t from (5) we get

$$T = t\left(1 + 2\frac{e}{1-e}\right) = \sqrt{\frac{2h}{g}}\left(\frac{1+e}{1-e}\right),$$

(Note: *Sum of infinite Geometrical series* $S_\infty = \frac{a}{1-r}$**).**

Note:

(i) If u be the velocity of the ball when it strikes the horizontal plane for the first time, then the velocities of successive rebounds will be eu, e^2u, e^3u, e^4u, ...

(ii) If h be the height of the ball where it falls freely on the horizontal plane for the first time. Then the heights attained by ball after successive rebounds will be e^2h, e^4h, e^6h, e^8h, ...

Example 2 A sphere A of mass 2*m* is moving on a smooth floor at right angle to a vertical wall, it collides directly on a stationary sphere B which is of mass 3*m*. After B rebounds from the wall, the spheres again collide directly. If *e* = 1/2 between A and B and *e* = 1/4 between B and wall, prove that the sphere B is brought to rest by the second impact of the spheres.

Solution 1st impact between A and B

Let the initial velocity of A be u, so that

$$u_1 = u, u_2 = 0, v_1 = v_A \text{ and } v_2 = v_B$$

By principle of **NEL** we have,

$$v_A - v_B = -e(u - 0)$$

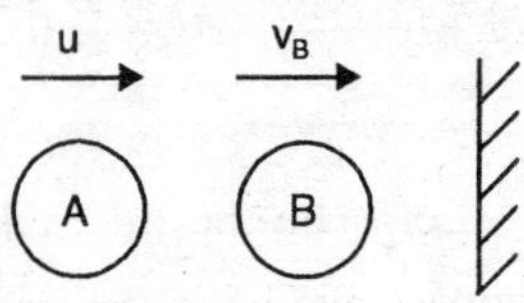

Fig. 1.27

Given $e = \frac{1}{2}$, between A and B.

$$\Rightarrow \quad v_A - v_B = -\frac{u}{2} \qquad \text{... (1)}$$

By principle of **LCM** we have,

$$2mv_A + 3mv_B = 2mu + 3m \times 0$$

$$\Rightarrow \quad 2v_A + 3v_B = 2u \qquad \text{... (2)}$$

On solving (1) and (2) we get,

$$v_A = \frac{u}{10},\ v_B = \frac{3u}{5} \qquad \text{... (3)}$$

Impact between B and wall

$$u_1 = \frac{3u}{5},\ u_2 = 0,\ v_1 = v_B' \text{ and } v_2 = 0$$

By principle of **NEL** we have,

$$v'_B - 0 = -e\left(\frac{3u}{5} - 0\right)$$

Given $e = \frac{1}{4}$, between B and the wall.

$$\Rightarrow \quad v'_B = -\frac{3u}{20} \qquad \ldots (4)$$

Therefore, B sphere will return with velocity $\frac{3u}{20}$.

2nd impact between A and B

$u_1 = v'_A = \frac{u}{10}$, $u_2 = v'_B = -\frac{3u}{20}$, $v_1 = v''_A$ and $v_2 = v''_B$

By principle of **NEL** we have,

$$v''_A - v''_B = -e\left\{\frac{u}{10} - \left(-\frac{3u}{20}\right)\right\}$$

Given $e = \frac{1}{2}$, between A and B.

$$\Rightarrow \quad v''_A - v''_B = -\frac{u}{8} \qquad \ldots (5)$$

By principle of **LCM** we have,

$$2mv''_A + 3mv''_B = 2m\frac{u}{10} + 3m \times -\frac{3u}{20}$$

$$\Rightarrow \quad 2v''_A + 3v''_B = -\frac{u}{4} \qquad \ldots (6)$$

On solving (5) and (6) we get,

$$\Rightarrow \quad v''_B = 0$$

Thus after the second impact between A and B, sphere B comes to rest.

§ 8. Oblique impact of a perfectly smooth sphere on a perfectly smooth fixed plane. Let the sphere strikes the perfectly smooth fixed plane at A. Let u and v be the velocities before and after impact, which make an angles α and θ with the normal to the plane at A (shown in fig).

(i) Resolution of velocities

The plane being perfectly smooth, there is no frictional force acting between the sphere and the plane along the tangent at A or perpendicular to normal

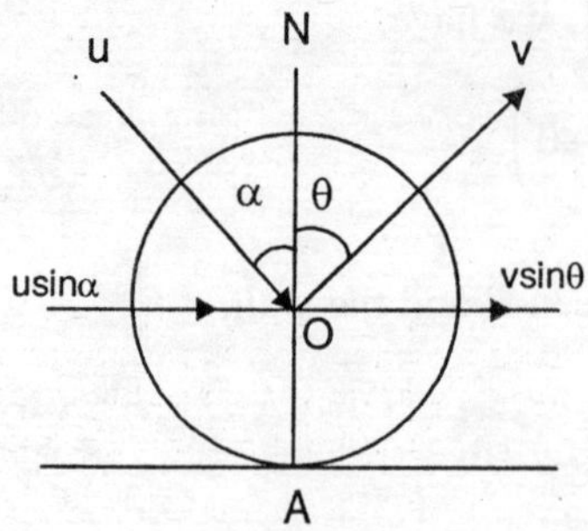

Fig. 1.28

AN. Hence the velocity of the sphere remains unchanged in the direction perpendicular to normal AN.

$$u\sin\alpha = v\sin\theta \qquad \text{... (i)}$$

(ii) By principle of NEL. Using NEL in the sense of normal AN, we have

$$v\cos\theta - 0 = -e(u\cos\alpha - 0)$$

$$\Rightarrow \quad v\cos\theta = -eu\cos\alpha \qquad \text{... (ii)}$$

Dividing (ii) by (i) we get,

$$\cot\theta = -e\cot\alpha \qquad \text{... (iii)}$$

Squaring and adding (i) and (ii) we get,

$$v^2 = u^2\left(\sin^2\alpha + e^2\cos^2\alpha\right). \qquad \text{... (iv)}$$

Loss of kinetic energy of the sphere due to impact $= \frac{1}{2}m\left(u^2 - v^2\right)$

$$= \frac{1}{2}mu^2(1-e^2)\cos^2\alpha.$$

impulse of blow on the surface = change of Momentum of the sphere along the normal AN

$$= mu\cos\alpha - mv\cos\theta = m(1+e)u\cos\alpha$$

Corollary 1. *When a perfectly elastic sphere strikes against a smooth fixed plane, its velocity remains unchanged in magnitude after impact and angle of incidence is equal to angle of reflection.*

Put $e = 1$, in (iii) and (iv) we get, $v = u$ and $\theta = \pi - \alpha$.

Corollary 2. *When a perfectly inelastic sphere strikes against a smooth fixed plane, the sphere slides along the plane with a velocity usina and does not bounce after impact.*

Put $e = 0$, in (iii) and (iv) we get, $v = u\sin\alpha$ and $\theta = \frac{\pi}{2}$.

Example 2 A ball is dropped vertically on a fixed smooth plane which is inclined at an angle of $\sin^{-1}\left(\frac{3}{5}\right)$ to the horizontal. If the ball strikes the plane with velocity v and rebounds in horizontal direction with velocity v_1, then prove that $e = \frac{9}{16}$ and $v_1 = \frac{3}{4}v$.

Solution Let $\sin^{-1}\left(\frac{3}{5}\right) = \theta \Rightarrow \sin\theta = \frac{3}{5}$, $\cos\theta = \frac{4}{5}$ and $\tan\theta = \frac{3}{4}$ ... (1)

(i) *Resolution of velocities* along the plane.

The plane being perfectly smooth, there is no frictional force acting between the ball and the plane along the plane. Hence the velocity of the ball remains unchanged along the plane

$$v\sin\theta = v_1 \sin(90° - \theta)$$

$\Rightarrow$ $v_1 = v\tan\theta$ ... (2)

In view of (1) and (2) we have

$\Rightarrow$ $v_1 = \frac{3}{4}v$ ... (3)

$$v_1 \cos(90^0 - \theta)$$

(ii) By principle of NEL. Using NEL in the sense of normal

$$v_1 \cos(90^0 - \theta) = -e(-v\cos\theta)$$

$\Rightarrow$ $v_1 \sin\theta = ev\cos\theta$

$\Rightarrow$ $v_1 \tan\theta = ev$ ... (4)

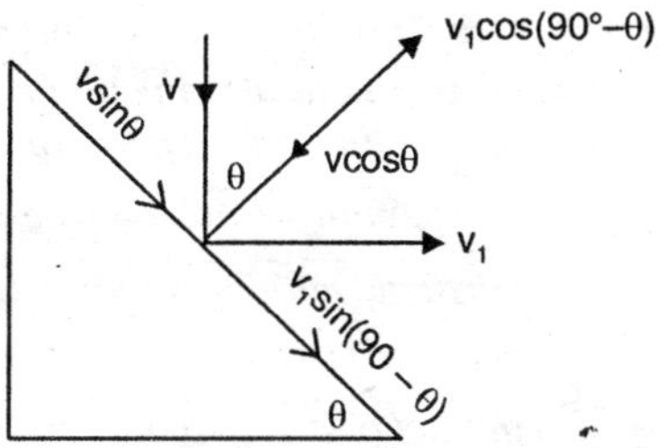

Fig. 1.29

In view of (1) and (4) we have

$\Rightarrow$ $v_1 \frac{3}{4} = ev$ ... (5)

In view of (3) and (5) we have $\frac{3}{4} \times \frac{3}{4} = e = \frac{9}{16}$

REVIEW EXERCISE 1.3

1. A ball falls from a height 64 cm on a smooth horizontal plane. If $e = \frac{1}{2}$, find the height to which it rises after rebounding four times.

(Ans. 0.25 cm)

2. A tennis ball is dropped from a point P at a height H above the surface of the court and it rebounds to the height h. Prove that the velocity V with which it must be projected vertically downward from P that it may just return to P after rebound is given by the equation

$$hV^2 = 2gH(H-h).$$

3. When a perfectly elastic sphere strikes against a smooth fixed plane. Show that its velocity remains unchanged in magnitude after impact and angle of incidence is equal to angle of reflection.

(**Hint.** Put $e = 1 \Rightarrow v = u$ and $\theta = \pi - \alpha$).

4. At what angle must a smooth sphere strike a horizontal plane so that after impact its direction may be at right angle to its former path.

(**Ans.** $\tan^{-1}\sqrt{e}$).

Note: *Readers are advised to see chapter-8 for any difficulty in forthcoming Miscellaneous Examples.*

MISCELLANEOUS EXAMPLES

Note: *For the impact of a body on the horizontal plane projected with velocity u and making an angle α with horizontal, following points are important:*

(*i*) *Consider the velocities along and perpendicular to the horizontal plane separately.*

(*ii*) *The velocity along the plane remains unchanged whereas the velocity perpendicular to the plane after each impact will be e times the velocity perpendicular to the plane before impact.*

(*iii*) *Horizontal velocity* $= u\cos\alpha$ *(constant through out the motion)*

(*iv*) *Vertical velocity* $= u\sin\alpha$ *(After each impact reduced by e times the velocity) i.e. Vertical velocity, after first impact* $= eu\sin\alpha$, *after second impact* $= e^2u\sin\alpha$ *and so on.*

(*v*) *Time of flight (T)* $= 2\times(\text{Vertical velocity})/g$,

$$Range = 2\times(Vertical\ velocity)\times\frac{Horizontal\ velocity}{g}$$

$$= (Horizontal\ velocity)T$$

Example 1 A particle is projected from a given point at the foot of one of the two smooth vertical walls. So as after three impacts it may return to the point of projection. If the last impact be direct, show that $e^3+e^2+e=1$.

Solution Let the particle be projected with components of velocity u along horizontal plane and v perpendicular to horizontal plane. Let B, C and D be the point of impacts on the wall. Let t_1, t_2, t_3, t_4, be the times for covering the distances AB, BC, DA respectively. And let AO = a

After striking B, C, D the horizontal components of velocities become eu, e^2u, e^3u respectively.

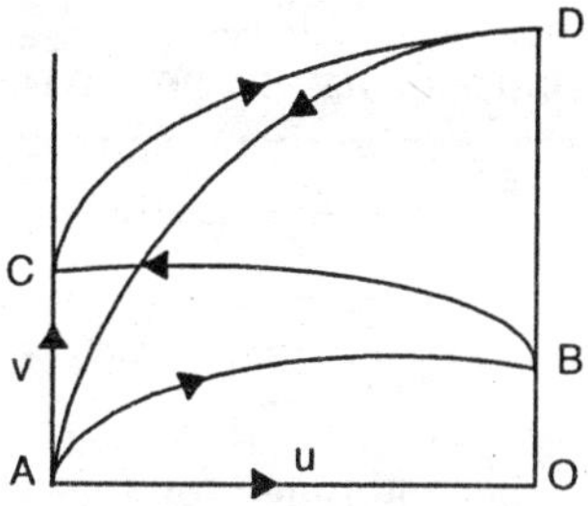

Fig. 1.30

$$t_1 = \frac{a}{u},\ t_2 = \frac{a}{eu},\ t_3 = \frac{a}{e^2u},\ t_4 = \frac{a}{e^3u} \quad \dots (1)$$

Since the impact at D is direct i.e. right angle to the wall, therefore, only horizontal component of the velocity will exist at D. Thus vertical component v will becomes zero after time $(t_1 + t_2 + t_3)$.

Moving from D to A it again retain the velocity v.

(**Note.** *In upward motion vertical component of velocity v gradually goes on decreasing due to gravity and in downward motion vertical component of velocity gradually goes on increasing from* 0 *to v due to gravity*)

$$\therefore\ t_1 + t_2 + t_3 = t_4 \qquad \dots (2)$$

In view of (1) and (2) we get,

$$\frac{a}{u} + \frac{a}{eu} + \frac{a}{e^2u} = \frac{a}{e^3u}$$

$$\Rightarrow \qquad 1 + \frac{1}{e} + \frac{1}{e^2} = \frac{1}{e^3}$$

$$\Rightarrow \qquad e^3 + e^2 + e = 1.$$

Example 2. A ball is projected from a point in a horizontal plane and it makes one rebound. Show that if the second range is equal to the greatest height reached by the ball, then the angle of projection will be $\tan^{-1} 4e$.

Solution Let u be the velocity of projection making the angle of α with the horizontal. The vertical components of velocity of first trajectory and second

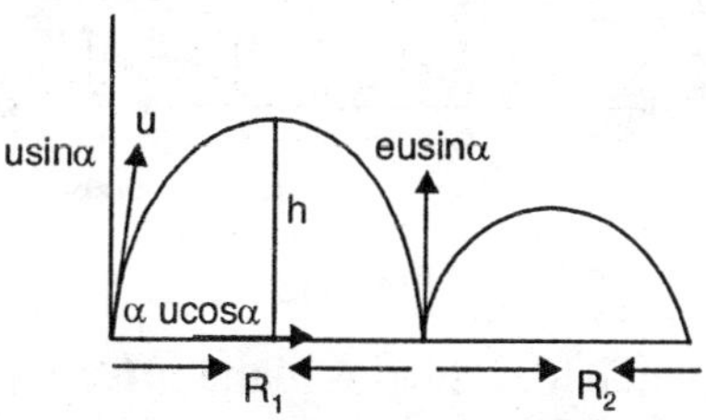

Fig. 1.31

trajectory will be $u\sin\alpha$ and $eu\sin\alpha$ respectively, **whereas the horizontal components of velocity remain constant for both** trajectories and equal to $u\cos\alpha$ throughout the motion.

Greatest height attained in first projectile be h.

$$\Rightarrow \quad h = \frac{u^2}{2g}\sin^2\alpha \qquad \ldots (1)$$

Second range for second projectile be R_2

$$\Rightarrow \quad R_2 = \frac{2(\text{Horizontal comp. of velocity}) \times (\text{Vertical comp. of velocity})}{g}$$

$$\Rightarrow \quad R_2 = \frac{2u\cos\alpha \times eu\sin\alpha}{g} = \frac{2u^2 e}{g}\sin\alpha\cos\alpha \qquad \ldots (2)$$

According to the problem, $R_2 = h$, thus in view of (1) and (2) we get,

$$\frac{2u^2 e}{g}\sin\alpha\cos\alpha = \frac{u^2}{2g}\sin^2\alpha$$

$$\Rightarrow \quad 4e\cos\alpha = \sin\alpha \Rightarrow \tan\alpha = 4e$$

$$\Rightarrow \quad \alpha = \tan^{-1} 4e \qquad \text{Hence proved}$$

Example 3 A ball is projected from a point in a horizontal plane with velocity u at an angle α to the horizontal. If e is the coefficient of restitution, show that it ceases to rebound from the plane at the end of time $\frac{2u\sin\alpha}{g(1-e)}$ and will have a range $\frac{u^2\sin\alpha}{g(1-e)}$.

Solution The ball will describe a parabolic path. Let the points of impact on horizontal plane be *A, B, C,* ...... (shown in figure). **The velocity along the plane remains unchanged after each impact** i.e. $u\cos\alpha$.

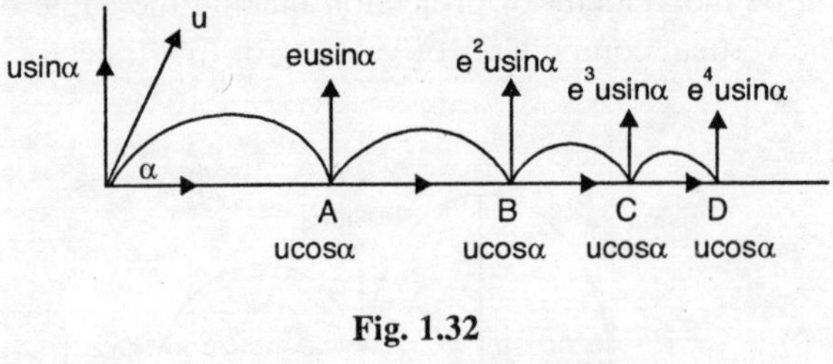

Fig. 1.32

The velocity perpendicular to the plane after each impact will be e times the velocity perpendicular to the plane before impact. Therefore,

Velocity perpendicular to the plane at $A = eu\sin\alpha$, Velocity perpendicular to the plane at $B = e^2u\sin\alpha$ Velocity perpendicular to the plane at $C = e^3u\ \sin\alpha$, and so on.

Time describing each projectile

$$= \text{Time of flight} = 2\times(\text{Vertical velocity})/g$$

Hence, time of describing projectile OA $= \dfrac{2u\sin\alpha}{g}$

Similarly, time of describing projectile AB $= \dfrac{2eu\sin\alpha}{g}$, time of describing projectile BC $= \dfrac{2e^2u\ \sin\alpha}{g}$ time of describing projectile CD $= \dfrac{2e^3u\sin\alpha}{g}$, and so on.

$$\therefore \quad \text{Total time}(T) = \frac{2u\sin\alpha}{g} + \frac{2eu\sin\alpha}{g} + \frac{2e^2u\sin\alpha}{g} + \frac{2e^3u\sin\alpha}{g} + \ldots$$

$$= \frac{2u\sin\alpha}{g}\left(1 + e + e^2 + e^3 + \ldots\infty\right)$$

(**Note.** *Sum of infinite Geometrical series* $S_\infty = \dfrac{a}{1-r}$)

$$\Rightarrow \qquad T = \frac{2u\sin\alpha}{g(1-e)}.$$

$$\text{Range}(R) = 2\times\left(\text{Vertical velocity}\right)\times(\text{Horizontal velocity})/g$$

$$= (\text{Horizontal velocity})\times T$$

$$\Rightarrow \qquad R = u\cos\alpha\times\frac{2u\sin\alpha}{g(1-e)} = \frac{2u^2\sin 2\alpha}{g(1-e)}.$$

Note. *For the impact of a body on the inclined plane projected with velocity u and making an angle α with the plane and plane is making an angle β following points are important:*

(*i*) *Consider the velocities along and perpendicular to the inclined plane separately.*

(*ii*) *velocity along the plane* $= u\cos\alpha$, *(Not affected due to impact)*

(*iii*) *velocity perpendicular to the plane* $= u\sin\alpha$ *Subject to retardation of* $g\cos\beta$, *for each projectile. And after each impact reduced by e times the velocity i.e. velocity perpendicular to the plane, after first impact* $= eu\sin\alpha$, *after second impact* $= e^2 u\sin\alpha$ *and so on.*

(*iv*) *Time of flight* $(T) = 2\times\dfrac{\text{Vertical velocity}}{g\cos\beta}$,

Range $= u\cos\alpha\times T - \dfrac{1}{2}g\sin\beta\times T^2$ *Subject to retardation of* $g\sin\beta$, *for each projectile.*

Example 4 A ball is projected from the foot of an inclined plane of inclination β with a velocity *u* at an angle α to the plane If *e* is the coefficient of restitution, show that it ceases to rebound from the plane at the end of time $\dfrac{2u\sin\alpha}{(1-e)g\cos\beta}$ and also find the total range along the plane.

Solution Components of velocity along and perpendicular to the plane are $u\cos\alpha$ and $u\sin\alpha$ respectively. For the first impact at A the initial velocity along the plane remains unchanged.

The velocity perpendicular to the plane after each impact will be *e* times the velocity perpendicular to the plane before impact. Therefore,

Velocity perpendicular to the plane at $A = eu\sin\alpha$, velocity perpendicular to the plane at $B = e^2 u\sin\alpha$, velocity perpendicular to the plane at $C = e^3 u\sin\alpha$, and so on (shown in fig).

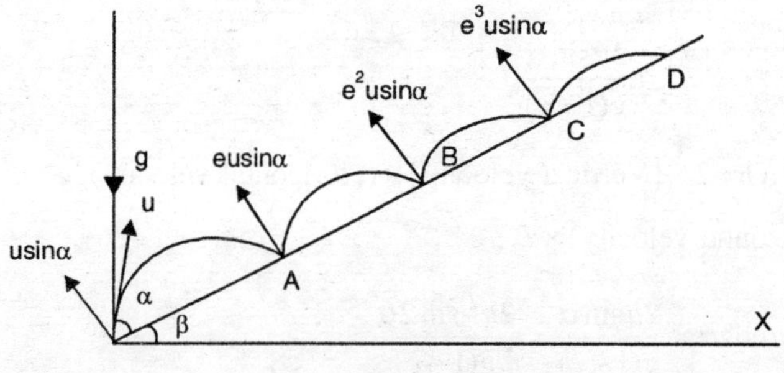

Fig. 1.33

We know that

$$\text{Time of flight } (T) = 2\times\frac{\text{Vertical velocity}}{g\cos\beta}$$

$$\text{Hence, time of describing projectile OA} = \frac{2u\sin\alpha}{g\cos\beta}$$

Similarly, time of describing projectile AB $=\dfrac{2eu\sin\alpha}{g\cos\beta}$, time of describing projectile BC $=\dfrac{2e^2u\sin\alpha}{g\cos\beta}$ time of describing projectile CD $=\dfrac{2e^3u\sin\alpha}{g\cos\beta}$, and so on.

$$\therefore \text{Total time}(T)=\frac{2u\sin\alpha}{g\cos\beta}+\frac{2eu\sin\alpha}{g\cos\beta}+\frac{2e^2u\sin\alpha}{g\cos\beta}+\frac{2e^3u\sin\alpha}{g\cos\beta}+\ldots$$

$$=\frac{2u\sin\alpha}{g\cos\beta}\left(1+e+e^2+e^3+\ldots\infty\right)$$

(**Note.** *Sum of infinite Geometrical series* $S_\infty=\dfrac{a}{1-r}$)

$$\Rightarrow \qquad T=\frac{2u\sin\alpha}{(1-e)g\cos\beta} \qquad \ldots (1)$$

Total range along the plane is equal to the distance covered by the ball with initial velocity $u\cos\alpha$ and subject to retardation of $g\sin\beta$. Thus we have,

$$\text{Total Range }(R)=u\cos\alpha\times T-\frac{1}{2}g\sin\beta\times T^2 \qquad \ldots (2)$$

In view of (1) and (2) we get,

$$R=T(u\cos\alpha-\frac{1}{2}g\sin\beta\times T)$$

$$=\frac{2u\sin\alpha}{(1-e)g\cos\beta}\left\{u\cos\alpha-\frac{1}{2}g\sin\beta\times\frac{2u\sin\alpha}{(1-e)g\cos\beta}\right\}$$

$$=\frac{2u^2\sin\alpha}{(1-e)^2g\cos^2\beta}\{(1-e)\cos\beta\cos\alpha-\sin\beta\sin\alpha\}$$

$$=\frac{2u^2\sin\alpha}{(1-e)^2g\cos^2\beta}\{\cos\beta\cos\alpha-\sin\beta\sin\alpha-e\cos\beta\cos\alpha\}$$

$$\Rightarrow \quad R=\frac{2u^2\sin\alpha}{(1-e)^2g\cos^2\beta}\{\cos(\alpha+\beta)-e\cos\beta\cos\alpha\}.$$

Note. *After travelling the distance R, there will be no more rebounds, however the ball will continue to move up the plane. If t be the time of moving up the* *plane and s be the total distance travelled along the plane with the effect of retardation of* $g\sin\beta$.

Using, $v = u + ft$ we get $0 = u\cos\alpha - g\sin\beta \times t \Rightarrow t = \dfrac{u\cos\alpha}{g\sin\beta}$.

Using, $s = ut + \dfrac{1}{2}ft^2$, we get,

$$s = u\cos\alpha\frac{u\cos\alpha}{g\sin\beta} - \frac{1}{2}g\sin\beta\left(\frac{u\cos\alpha}{g\sin\beta}\right)^2 \Rightarrow s = \frac{u^2\cos^2\alpha}{2g\sin\beta}.$$

Example 4 A ball is projected with a velocity *u* from the point on the inclined plane of inclination α to the horizontal, the direction of projection at right angle to the plane. Prove that before ceasing to rebound, it will describe a length $\dfrac{2u^2\sin\alpha}{(1-e)^2 g\cos^2\alpha}$. Where *e* is the coefficient of restitution.

Solution Let the ball be projected from O point on the plane with a velocity *u* right angle to the inclined plane Let the points of impact on horizontal plane be *A, B, C,* ...... (shown in figure).

Then the velocity components perpendicular to the plane after impacts at *A, B, C,* ...... will be eu, e^2u, e^3u ......etc.

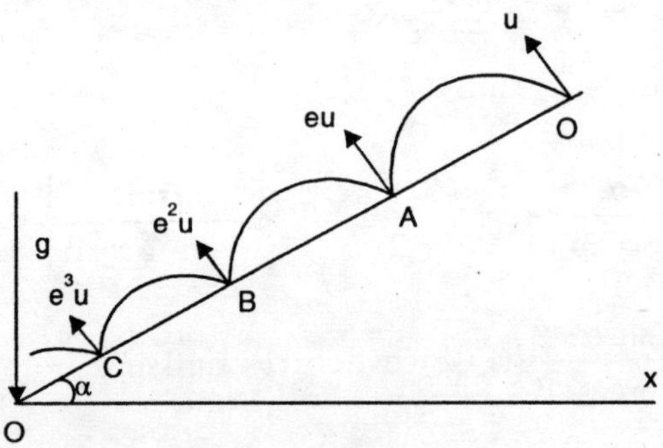

Fig. 1.34

Time describing each projectile

$$= \text{Time of flight} = 2 \times (\text{Vertical velocity})/g\cos\alpha$$

Hence, time of describing projectile OA $= \dfrac{2u}{g\cos\alpha}$

Similarly, time of describing projectile AB $= \dfrac{2eu}{g\cos\alpha}$, time of describing projectile BC $= \dfrac{2e^2u}{g\cos\alpha}$, and so on.

$$\therefore \quad \text{Total time } (T) = \frac{2u}{g\cos\alpha} + \frac{2eu}{g\cos\alpha} + \frac{2e^2u}{g\cos\alpha} + \ldots$$

$$= \frac{2u}{g\cos\alpha}\left(1 + e + e^2 + e^3 + \ldots\infty\right)$$

(**Note.** *Sum of infinite Geometrical series* $S_\infty = \dfrac{a}{1-r}$)

$$\Rightarrow \qquad T = \frac{2u}{(1-e)g\cos\alpha} \qquad \ldots (1)$$

Initial velocity along the plane = 0 with the acceleration of $g\sin\alpha$, now using $s = ut + \frac{1}{2}ft^2$ and in view of (1), we get,

$$\text{Total length } (R) = 0 + \frac{1}{2}g\sin\alpha\left(\frac{2u}{(1-e)g\cos\alpha}\right)^2$$

$$\Rightarrow \qquad R = \frac{2u^2\sin\alpha}{(1-e)^2 g\cos^2\alpha}.$$

REVIEW EXERCISE 1.4

1. A ball is projected from a point in a horizontal plane; after one rebound from the plane it strikes directly against a vertical wall, after two more rebounds it return to the point of projection. Prove that $2e^3 + e - 2 = 0$.
2. From a point on a smooth horizontal plane a ball is projected with a velocity u at an angle α to the horizontal. Show that it will keep rebounding from the plane for the time $\dfrac{2u\sin\alpha}{g(1-e)}$, where e be the coefficient of restitution.
3. A particle is projected at an angle α to the horizontal to strike a smooth vertical wall and after rebounding it passes through the point of projection. If Φ be the angle of inclination to the horizontal at which the rebounds from the wall if e be the coefficient of restitution, show that $e(1 + e)\tan\Phi = (1-e)\tan\alpha$.

4. A particle is projected from the foot of a plane of inclination α with the horizontal, in the direction making an angle θ with the plane and rebounds vertically at the first impact. Show that $\cot\theta = (2+e)\tan\alpha$, where e is the coefficient of restitution.
5. A particle is dropped from a vertical height a upon the highest point of a plane, which is of length b and is inclined at an angle α to the horizontal and descends to the bottom in the three jumps, prove that $b = 4ae\,(1 + e)(1+e+e^2)(1+e^2)\sin\alpha$. Where e is the coefficient of restitution.

2 Motion in Two Dimension in Polar Coordinates

§ 1. Radial and Transverse Components of Velocity

Consider a particle be at $P(r, \theta)$ moving on a plane curve $r = f(\theta)$, referred to O as pole and OX as initial line. Let this position $P(r, \theta)$ be particle at any instant t. Let the particle after time $(t + \delta t)$ move to a neighboring point $Q(r + \delta r, \theta + \delta\theta)$.

Draw a perpendicular QR from Q on OP.

We have OQ $= r + \delta r$, OP $= r$

$$QR = (r + \delta r)\sin\delta\theta, \; OR = (r + \delta r)\cos\delta\theta$$

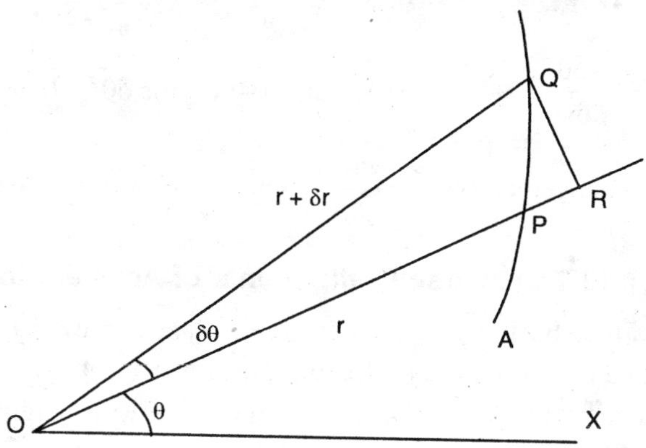

Fig. 2.1

Radial velocity means the rate of change of displacement along radial direction during the interval δt. As the particle moves from P to Q clearly, the displacement along radial direction (along OP) is $PR = OR - OP$.

Radial velocity (**u**)

$$= \text{Lt}_{\delta t \to 0} \frac{\text{Change of displacemental on gradial direction (along OP)}}{\delta t}$$

$$= \text{Lt}_{\delta t \to 0} \frac{PR}{\delta t}$$

$$= \text{Lt}_{\delta t \to 0} \frac{OR - OP}{\delta t} = \text{Lt}_{\delta t \to 0} \frac{(r + \delta r)\cos \delta\theta - r}{\delta t} = \text{Lt}_{\delta t \to 0} \frac{\delta r}{\delta t}$$

(As $\cos \delta\theta \to 1$ for $\delta\theta \to 0$)

$\Rightarrow$ Radial velocity $(u) = \dfrac{dr}{dt} = \dot{r}$, in the direction of r increasing.

Transverse velocity means the rate of change of displacement perpendicular to radius vector during the interval δt.

$$\text{Transverse velocity } (v) = \text{Lt}_{\delta t \to 0} \frac{\text{Change of displacement perpendicular to radial direction (perp. to OP)}}{\delta t}$$

$$= \text{Lt}_{\delta t \to 0} \frac{QR}{\delta t}$$

$$= \text{Lt}_{\delta t \to 0} \frac{(r + \delta r)\sin \delta\theta}{\delta t} \text{Lt}_{\delta t \to 0} \frac{r\delta\theta}{\delta t}$$

$$\text{Lt}_{\delta t \to 0} \frac{r\delta\theta}{\delta t}$$

(As $\sin \delta\theta \to \delta\theta$ for $\delta\theta \to 0$ and $\delta r \delta\theta \to 0$)

$\Rightarrow$ Transverse velocity $(v) = r\dfrac{d\theta}{dt} = r\dot{\theta}$ in sense of θ increasing.

§ 2. Radial and Transverse Components of Acceleration

Consider a particle be at $P(r, \theta)$ moving on a plane curve $r = f(\theta)$, referred to O as pole and OX as initial line. Let this position $P(r, \theta)$ be particle at any instant t. Let the particle after time $(t + \delta t)$ move to a neighboring point $Q(r + \delta r, \theta + \delta\theta)$.

Now let the components of velocity along and Q perpendicular to OP be u and v respectively and the components of velocity along and perpendicular to OQ be $u + \delta u$ and $v + \delta v$ respectively.

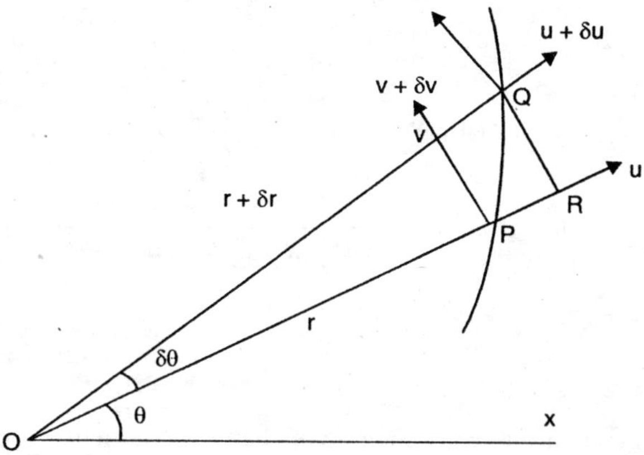

Fig. 2.2

Radial Acceleration: Radial Acceleration means the rate of change of velocity along radial direction during the interval δt.

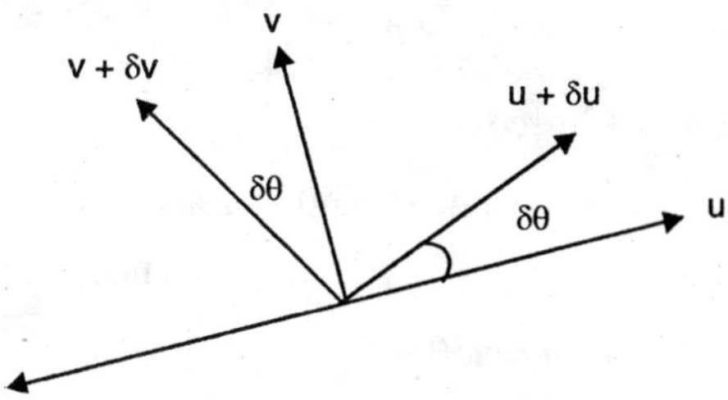

Fig. 2.3

Thus the change of velocity along radial direction during interval δt = Resolution of velocity components at Q along radial direction

$-$Radial velocity at P

$= (u+\delta u)\cos\delta\theta - (v+\delta v)\sin\delta\theta - u$

(As $\cos\delta\theta \to 1$ and $\sin\delta\theta \to \delta\theta$ for $\delta\theta \to 0$)

neglecting the term $\delta v \delta\theta$.

$$= (u+\delta u) - (v+\delta v)\delta\theta - u = \delta u - v\delta\theta \quad \ldots \text{(i)}$$

Radial Acceleration

$$= \mathrm{Lt}_{\delta t \to 0} \frac{\text{Change of velocity along radial direction (along OP)}}{\delta t}$$

$$= Lt_{\delta t \to 0} \frac{\delta u - v\delta\theta}{\delta t} = \frac{du}{dt} - \frac{d\theta}{dt} \ldots \text{(ii)}$$

We know that radial velocity $u = \frac{dr}{dt} = \dot{r}$ and transverse velocity $v = r\frac{d\theta}{dt} = r\dot{\theta}$, putting in (ii), we get;

$$\Rightarrow \quad \text{Radial Acceleration} = (f_r) = \frac{d}{dt}\left(\frac{dr}{dt}\right) - r\frac{d\theta}{dt}\frac{d\theta}{dt} = \frac{d^2r}{dt^2} - r\left(\frac{d\theta}{dt}\right)^2$$

$$\Rightarrow \quad f_r = \ddot{r} - r\dot{\theta}^2.$$

Transverse acceleration

Transverse acceleration means the rate of change of velocity perpendicular to radial direction during the interval δt.

Thus the change of velocity perpendicular to radial direction during interval δt

= *Resolution of velocity components at Q perpendicular to radial direction – Transverse velocity at P*

$$= (u + \delta u)\sin\delta\theta + (v + \delta v)\cos\delta\theta - v$$

(As $\cos\delta\theta \to 1$ and $\sin\delta\theta \to \delta\theta$ for $\delta\theta \to 0$)

neglecting the term $\delta u \delta\theta$

$$= (u + \delta u)\delta\theta + (v + \delta v) - u = u\delta\theta + \delta v \qquad \text{... (iii)}$$

$$\text{Transverse Acceleration} = \text{Lt}_{\delta t \to 0} \frac{\text{Change of velocity perpendicular to radial direction (along OP)}}{\delta t}$$

$$= \text{Lt}_{\delta t \to 0} \frac{u\delta\theta + \delta v}{\delta t} = u\frac{d\theta}{dt} + \frac{dv}{dt} \qquad \text{... (iv)}$$

We know that radial velocity $u = \frac{dr}{dt} = \dot{r}$ and transverse velocity $v = r\frac{d\theta}{dt} = r\dot{\theta}$, putting in (iv), we get;

$$\Rightarrow \quad \text{Transverse Acceleration}\,(f_t) = \frac{dr}{dt}\frac{d\theta}{dt} + \frac{d}{dt}\left(r\frac{d\theta}{dt}\right) = 2\frac{dr}{dt}\frac{d\theta}{dt} + r\frac{d^2\theta}{dt^2}$$

$$\Rightarrow \quad f_t = r\ddot{\theta} + 2\dot{r}\dot{\theta} = \frac{1}{r}\frac{d(r^2\dot{\theta})}{dt}.$$

Alternative method for Radial and Transverse Components of Velocity and Acceleration

The velocities and accelerations of a particle at P (x, y) parallel to x and y axes are $\dot{x}$, $\dot{y}$ and $\ddot{x}$, $\ddot{y}$ respectively. If (x, y) and (r, θ) be the coordinates of point P in Cartesian coordinate and polar coordinates respectively. Then we know that

$$x = r\cos\theta,\ y = r\sin\theta \qquad \text{... (i)}$$

Differentiating (i) w.r.t. t we get

$$\dot{x} = \dot{r}\cos\theta - (r\sin\theta)\dot{\theta} \qquad \text{... (ii)}$$

$$\dot{y} = \dot{r}\sin\theta + (r\cos\theta)\dot{\theta} \qquad \text{... (iii)}$$

Differentiating again (ii) and (iii) w.r.t. t we get,

$$\ddot{x} = \ddot{r}\cos\theta - 2\dot{r}\dot{\theta}\sin\theta - (r\cos\theta)\dot{\theta}^2 - (r\sin\theta)\ddot{\theta} \qquad \text{... (iv)}$$

$$\ddot{y} = \ddot{r}\sin\theta + 2\dot{r}\dot{\theta}\cos\theta - (r\sin\theta)\dot{\theta}^2 + (r\cos\theta)\ddot{\theta} \qquad \text{... (v)}$$

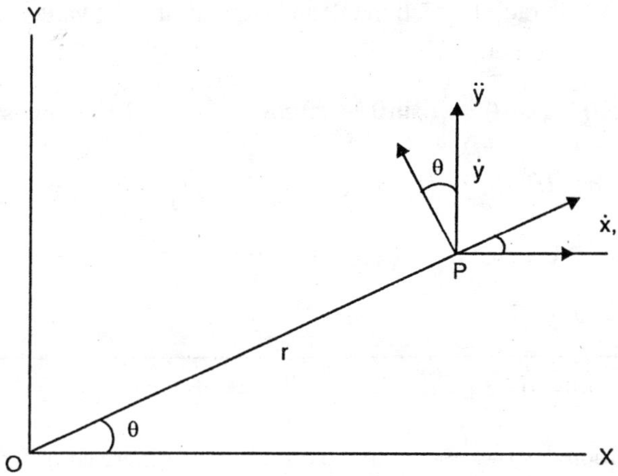

Fig. 2.4

Radial velocity (*u*) – Now resolving velocities $\dot{x}$ and $\dot{y}$ in the radial direction OP and substituting the values of $\dot{x}$ and $\dot{y}$ from (ii) and (iii) we get,

$$u = \dot{x}\cos\theta + \dot{y}\sin\theta = \left\{\dot{r}\cos\theta - (r\sin\theta)\dot{\theta}\right\}\cos\theta + \left\{\dot{r}\sin\theta + (r\cos\theta)\dot{\theta}\right\}\sin\theta$$

$$\Rightarrow \quad u = \dot{r}(\cos^2\theta + \sin^2\theta) = \dot{r}$$

Transverse velocity (*v*) – Now resolving velocities $\dot{x}$ and $\dot{y}$ perpendicular to the radial direction OP and substituting the values of $\dot{x}$ and $\dot{y}$ from (ii) and (iii) we get,

$$v = \dot{y}\cos\theta - \dot{x}\sin\theta = \left\{\dot{r}\sin\theta + (r\cos\theta)\dot{\theta}\right\}\cos\theta - \left\{\dot{r}\cos\theta - (r\sin\theta)\dot{\theta}\right\}\sin\theta$$

$$\Rightarrow \; v = r\dot{\theta}(\cos^2\theta + \sin^2\theta) = r\dot{\theta}$$

Radial Acceleration (f_r) – Now resolving accelerations $\ddot{x}$ *and* $\ddot{y}$ in the radial direction OP and substituting the values of $\ddot{x}$ *and* $\ddot{y}$ from (iv) and (v) we get,

$$f_r = \ddot{x}\cos\theta + \ddot{y}\sin\theta = \left\{\ddot{r}\cos\theta - 2\dot{r}\dot{\theta}\sin\theta - (r\cos\theta)\dot{\theta}^2 - (r\sin\theta)\ddot{\theta}\right\}\cos\theta$$

$$+\left\{\ddot{r}\sin\theta + 2\dot{r}\dot{\theta}\cos\theta - (r\sin\theta)\dot{\theta}^2 + (r\cos\theta)\ddot{\theta}\right\}\sin\theta$$

$$\Rightarrow \quad f_r = \ddot{r}(\cos^2\theta + \sin^2\theta) - r\dot{\theta}^2(\cos^2\theta + \sin^2\theta)$$

$$\Rightarrow \quad f_r = \ddot{r} - r\dot{\theta}^2$$

Transverse Acceleration (f_t) – Now resolving accelerations $\ddot{x}$ and $\ddot{y}$ perpendicular to the radial direction OP and substituting the values of $\ddot{x}$ and $\ddot{y}$ from (iv) and (v) we get,

$$f_t = \ddot{y}\cos\theta - \ddot{x}\sin\theta = \left\{\ddot{r}\sin\theta + 2\dot{r}\dot{\theta}\cos\theta - (r\sin\theta)\dot{\theta}^2 + (r\cos\theta)\ddot{\theta}\right\}\cos\theta$$

$$-\left\{\ddot{r}\cos\theta - 2\dot{r}\dot{\theta}\sin\theta - (r\cos\theta)\dot{\theta}^2 - (r\sin\theta)\ddot{\theta}\right\}\sin\theta$$

$$\Rightarrow \; f_t = 2\dot{r}\dot{\theta}(\cos^2\theta + \sin^2\theta) + r\ddot{\theta}(\cos^2\theta + \sin^2\theta)$$

$$\Rightarrow \; f_t = r\ddot{\theta} + 2\dot{r}\dot{\theta}$$

Important Results to be used

1. Radial velocity $(u) = \dfrac{dr}{dt} = \dot{r}$, in the direction of r increasing, Transverse velocity $(v) = r\dfrac{d\theta}{dt} = r\dot{\theta}$ in sense of θ increasing.
2. Radial Acceleration $(f_r) = \dfrac{d^2r}{dt^2} - r\left(\dfrac{d\theta}{dt}\right)^2 = \ddot{r} - r\dot{\theta}^2$, Transverse Acceleration $(f_t) = 2\dfrac{dr}{dt}\dfrac{d\theta}{dt} + r\dfrac{d^2\theta}{dt^2} = r\ddot{\theta} + 2\dot{r}\dot{\theta} = \dfrac{1}{r}\dfrac{d\left(r^2\dot{\theta}\right)}{dt}$
3. Resultant velocity $\vec{v} = \dot{r}\hat{r} + r\dot{\theta}\hat{s}$, here $\hat{r}$ and $\hat{s}$ are the unit vectors along and perpendicular to radius vector respectively. $\vec{v}$

Will be along the tangent to the curve at that point and

$|\vec{v}| = \sqrt{(\dot{r})^2 + (r\dot{\theta})^2}$.

4. $\frac{d}{dt}\hat{r} = \omega\hat{s}, \frac{d}{dt}\hat{s} = -\omega\hat{s}$ where $\frac{d\theta}{dt} = \omega$.

ILLUSTRATIVE EXAMPLES

Example 1 If the radial and transverse speeds of a particle are always proportional to each other, show that its path is an equiangular spiral $r = ae^{k\theta}$.

Solution Here we have

Radial velocity $(u) \propto$ Transverse velocity (v)

$\Rightarrow \quad u = kv$, where k is a constant of proportionality.

$$\Rightarrow \quad \frac{dr}{dt} = kr\frac{d\theta}{dt}$$

$\Rightarrow \quad \frac{dr}{r} = kd\theta$, on integration we get,

$$\Rightarrow \quad \log r = k\theta + C$$

$\Rightarrow \quad r = ae^{k\theta}$ where $C = \log a$.

$r = ae^{k\theta}$ is the equation of an equiangular spiral.

Example 2 If the radial and transverse speeds of a particle with respect to fixed origin are λr^2 and $\mu\theta^2$ respectively. Find the path of the particle and magnitude of radial and transverse acceleration.

Solution Given:

Radial velocity $(u) = \lambda r^2$ and Transverse velocity $(v) = \mu\theta^2$

$$\Rightarrow \quad \dot{r} = \frac{dr}{dt} = \lambda r^2 \qquad \ldots (1)$$

$$\Rightarrow \quad r\dot{\theta} = r\frac{d\theta}{dt} = \mu\theta^2 \qquad \ldots (2)$$

Dividing (1) by (2) we get

$\frac{dr}{rd\theta} = \frac{\lambda r^2}{\mu\theta^2} \Rightarrow \frac{dr}{r^3} = \frac{\lambda d\theta}{\mu\theta^2}$, on integration we get, $\frac{-1}{2r^3} = -\frac{\lambda}{\mu\theta} + C'$

$$\Rightarrow \frac{1}{2r^3} = \frac{\lambda}{\mu\theta} + C, \text{ required equation of the path traced by the particle.}$$

Radial Acceleration $(f_r) = \ddot{r} - r\dot{\theta}^2 = \frac{d^2r}{dt^2} - r\left(\frac{d\theta}{dt}\right)^2$

$$\Rightarrow \quad (f_r) = \frac{d}{dt}\left(\frac{dr}{dt}\right) - \frac{1}{r}\left(r\frac{d\theta}{dt}\right)^2 \qquad \text{... (3)}$$

In view of (1) and (2), (3) reduces to

$$(f_r) = \frac{d}{dt}\lambda r^2 - \frac{1}{r}\left(\mu\theta^2\right)^2$$

$$\Rightarrow \quad (f_r) = 2\lambda r\frac{dr}{dt} - \frac{1}{r}\mu^2\theta^4 \qquad \left(\text{using (1) i.e., } \frac{dr}{dt} = \lambda r^2\right)$$

$$\Rightarrow \quad (f_r) = 2\lambda^2 r^3 - \frac{1}{r}\mu^2\theta^4, \qquad \text{required expression for } f_r.$$

Transverse Acceleration $(f_t) = r\ddot{\theta} + 2\dot{r}\dot{\theta} = \frac{1}{r}\frac{d\left(r^2\dot{\theta}\right)}{dt} = \frac{1}{r}\frac{d\left(r r\dot{\theta}\right)}{dt}$,

(using (2) i.e. $r\dot{\theta} = \mu\theta^2$)

$$\Rightarrow \quad f_t = \frac{1}{r}\frac{d}{dt}\left(r\mu\theta^2\right) = \frac{\mu}{r}\left\{\theta^2\frac{dr}{dt} + 2r\theta\frac{d\theta}{dt}\right\} \qquad \text{((using (1) and (2))}$$

$$\Rightarrow \quad f_t = \frac{\mu}{r}\left\{\theta^2\lambda r^2 + 2\mu\theta^2\theta\right\}$$

$$\Rightarrow \quad f_t = \lambda\mu r\theta^2 + \frac{2\mu^2\theta^3}{r}.$$

Example 3 If the path of particle is $r = a\tan\theta$ and the acceleration is only directed towards the origin. Show that the acceleration is

$$\frac{\mu^2}{r^3}\left(3 + \frac{2a^2}{r^2}\right) \text{ where } \mu = r^2\frac{d\theta}{dt}.$$

Solution The acceleration is only directed towards the origin implies transverse acceleration (f_t) is zero.

$$\Rightarrow \quad f_t = \frac{1}{r}\frac{d\left(r^2\dot{\theta}\right)}{dt} = 0$$

$$\Rightarrow \quad r^2\dot{\theta} = r^2\frac{d\theta}{dt} = \mu \qquad \text{... (1) where } \mu \text{, is constant}$$

$$\text{Radial Acceleration } (f_r) = \ddot{r} - r\dot{\theta}^2 = \frac{d^2r}{dt^2} - r\left(\frac{d\theta}{dt}\right)^2$$

We have $r = a\tan\theta$ ... (2)

$$\Rightarrow \quad \frac{dr}{dt} = a\sec^2\theta\frac{d\theta}{dt} \qquad \text{(using (1) and (2))}$$

$$\Rightarrow \quad \frac{dr}{dt} = a\sec^2\theta\frac{\mu}{r^2} = a\left(1+\tan^2\theta\right)\frac{\mu}{r^2} = a(1+\frac{r^2}{a^2})\frac{\mu}{r^2}$$

$$\Rightarrow \quad \frac{dr}{dt} = a\frac{\mu}{r^2} + \frac{\mu}{a} \qquad \text{... (3)}$$

$$\Rightarrow \quad \frac{d^2r}{dt^2} = \frac{d}{dt}\frac{dr}{dt} = \frac{d}{dt}\left(a\frac{\mu}{r^2} + \frac{\mu}{a}\right) = -2a\frac{\mu}{r^3}\frac{dr}{dt} \qquad \text{(using (3))}$$

$$= -2a\frac{\mu}{r^3}\left(a\frac{\mu}{r^2} + \frac{\mu}{a}\right) = -2\frac{\mu^2}{r^3}\left(\frac{a^2}{r^2} + 1\right)$$

$$\Rightarrow \quad \frac{d^2r}{dt^2} = -2\frac{\mu^2}{r^3}\left(\frac{a^2}{r^2} + 1\right) \qquad \text{... (4)}$$

$$f_r = \ddot{r} - r\dot{\theta}^2 = \frac{d^2r}{dt^2} - r\left(\frac{d\theta}{dt}\right)^2 \qquad \text{(using (1) and (4))}$$

$$\Rightarrow \quad f_r = -2\frac{\mu^2}{r^3}\left(\frac{a^2}{r^2} + 1\right) - r\left(\frac{\mu}{r^2}\right)^2$$

$$\Rightarrow \quad f_r = -\frac{\mu^2}{r^3}\left(3 + \frac{2a^2}{r^2}\right)$$

The radial acceleration is itself negative. Thus, the acceleration is directed towards the origin because it is in the sense of r decreasing.

Example 4 If a particle describes an equiangular spiral $r = ae^{m\theta}$ with constant velocity. Find the radial and transverse components of acceleration.

Solution We have

$$|\vec{v}| = \sqrt{(\dot{r})^2 + (r\dot{\theta})^2} = k \qquad \dots (1)$$

where k is a constant $r = ae^{m\theta}$

$$\Rightarrow \quad \dot{r} = \frac{dr}{dt} = mae^{m\theta}\frac{d\theta}{dt} \qquad \Rightarrow \quad \dot{r} = \frac{dr}{dt} = mr\dot{\theta} \qquad \dots (2)$$

Substituting the value of $\dot{r}$ from (2) in (1), we get,

$$k = \sqrt{(\dot{r})^2 + (r\dot{\theta})^2} = \sqrt{(mr\dot{\theta}) + (r\dot{\theta})^2} = r\dot{\theta}\sqrt{m^2+1}$$

$$\Rightarrow \quad r\dot{\theta} = \frac{k}{\sqrt{m^2+1}} = \text{constant} \qquad \dots (3)$$

Substituting the value of $r\dot{\theta}$ from (3) in (2), we get,

$$\dot{r} = \frac{dr}{dt} = m\frac{k}{\sqrt{m^2+1}} = \text{constant} \qquad \dots (4)$$

$$\ddot{r} = \frac{d}{dt}\frac{dr}{dt} = \frac{d}{dt}\left(m\frac{k}{\sqrt{m^2+1}}\right) = 0 \qquad \dots (5)$$

$$\text{Radial Acceleration}\,(f_r) = \ddot{r} - r\dot{\theta}^2 = 0 - r\left(\frac{1}{r}\frac{k}{\sqrt{m^2+1}}\right)^2 = -\frac{1}{r}\frac{k^2}{(m^2+1)}.$$

$$\text{Transverse Acceleration }(f_t) = r\ddot{\theta} + 2\dot{r}\dot{\theta} = \frac{1}{r}\frac{d(rr\dot{\theta})}{dt} = \frac{1}{r}\frac{d}{dt}\left(r\frac{k}{\sqrt{m^2+1}}\right)$$

(using (4))

$$\Rightarrow \quad f_t = \frac{k}{r\sqrt{m^2+1}}\dot{r} = \frac{k}{r\sqrt{m^2+1}}m\frac{k}{\sqrt{m^2+1}}\frac{m}{r}\frac{k^2}{(m^2+1)}$$

REVIEW EXERCISE 2.1

1. If the angular velocity of a moving point about the fixed origin be constant, show that its transverse acceleration directly proportional to radial velocity. **(Hint.** $\theta = k$ (constant)
2. If the radial and transverse speeds of a particle with respect to fixed origin are λr and $\mu\theta$ respectively. Find the path of the particle and magnitude of radial and transverse accelerations.

(**Ans.** Path $-\frac{1}{r} = \frac{\lambda}{\mu}\log\theta + C$, $(f_r) = \lambda^2 r - \frac{1}{r}\mu^2\theta^2$ and $f_t = \mu\theta\left(\lambda + \frac{\mu}{r}\right)$

3. If a particle describes an equiangular spiral $r = ae^{\theta}$ such that it has no radial component of acceleration. Prove that it will have constant angular velocity and magnitude of both velocity and acceleration directly proportional to r.
4. If a particle describes an equiangular spiral $r = ae^{\theta}$ such that it has constant angular velocity. Prove that it will have no radial component acceleration and magnitude of transverse acceleration directly proportional to r.
5. If the radial and transverse speeds of a particle are always proportional to each other and same behavior with the radial and transverse accelerations. Show that the velocity of the particle varies as some power of the radius vector.
6. A particle moves with a constant angular speed ω about O in a plane. If the rate of change of acceleration parallel to PO, then $\frac{d^2r}{dt^2} = \frac{\omega^2}{3}r$.

   (Hint: $\frac{d}{dt}\hat{r} = \omega\hat{s}$, $\frac{d}{dt}\hat{s} = -\omega\hat{s}$, $\frac{d\theta}{dt} = \omega$ and $\vec{f} = f_r\hat{r} + f_t\hat{s}\frac{d}{dt}\vec{f}$, is parallel to PO $\Rightarrow$ Coefficient of $\hat{s}$ in $\frac{d}{dt}\vec{f}$ is zero)
7. A particle moves along a circle $r = 2a\cos\theta$ in such a way that the acceleration towards the origin is always zero. Prove that

$$\frac{d^2\theta}{dt^2} = -2\cot\theta\,\dot{\theta}^2.$$

§ 3. Pedal Equation (*p* – *r* equation)

Let the equation of curve in polar coordinates be $r = f(\theta)$, r be the radius vector of an arbitrary point P and p be the length of perpendicular from origin upon the tangent at point P. **Then the equation of curve in terms of *p* and *r* is known as pedal equation (*p* – *r* equation) of the curve.** In pedal equation p and r are connected by the following relation: $\frac{1}{p^2} = \frac{1}{r^2} + \frac{1}{r^4}\left(\frac{dr}{d\theta}\right)^2$ and

$\tan\Phi = r\frac{d\theta}{dr}$ Φ, be the angle between the radius vector and the tangent.

Proof. Let O be the origin, fixed line OX as initial line $\vec{r}$, be the position vector of P and let the particle be at point $P(r, \theta)$ on the curve $r = f(\theta)$ at time t. Let $\vec{v}$ be the velocity vector of the particle along the tangent at P, $OM = p$ be the length of perpendicular from O upon the tangent O and ϕ *be* the angle between the radius vector and the tangent.

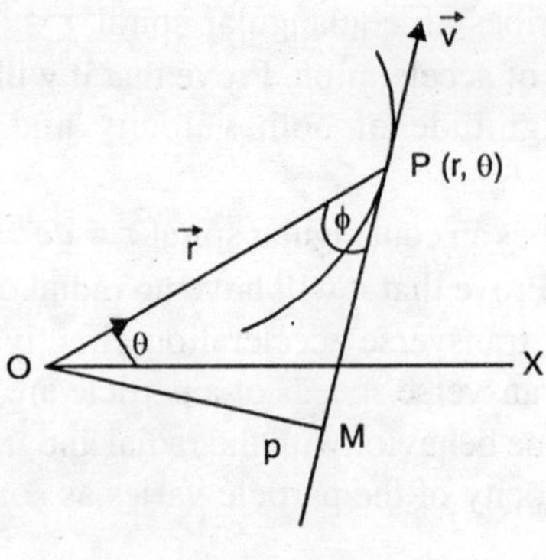

Fig. 2.5

In ΔOMP we have

$$p = r\sin\phi = \frac{r|\vec{v}|\sin\phi}{|\vec{v}|} \quad \text{... (i)}$$

$[\vec{r} = |\vec{r}|\,\hat{r} = r\hat{r}$ *as* $|\vec{r}| = r =$ length of radius vector$]$

$= |\vec{r} \times \hat{v}|$, $\quad [\,\hat{v} = \dfrac{\vec{v}}{|\vec{v}|}$ = unit velocity vector and vector product

$|\vec{a} \times \hat{b}| = |\vec{a}|\, sin\theta,\ |\hat{b}| = 1]$

$$= \frac{1}{|\vec{v}|}|\vec{r} \times \vec{v}|$$

$$= \frac{1}{|\vec{v}|}|\vec{r} \times (\dot{r}\hat{r} + r\dot{\theta}\hat{s})| \quad \text{... (ii)}$$

$[\,\vec{v} = (\dot{r}\hat{r} + r\dot{\theta}\hat{s})$ and $\hat{s})$ are resp. unit vectors along and perpendicular to radius vector].

$= \dfrac{1}{|\vec{v}|}|\vec{r} \times r\dot{\theta}\hat{s}|$ $[\,|\vec{r} \times \hat{r}| = 0$, as both the vectors having same direction]

$$= \frac{1}{|\vec{v}|}|r\hat{r} \times r\dot{\theta}\hat{s}|$$

$= \dfrac{r^2\dot{\theta}}{|\vec{v}|}|\hat{r} \times \hat{s}|$ using $|\hat{r} \times \hat{s}| = 1$,

[as the vector product of two unit and mutually perpendicular vectors is unit vector]

$$\Rightarrow \quad p = \frac{r^2\dot{\theta}}{|\vec{v}|}$$

$$\Rightarrow \quad \frac{1}{p^2} = \frac{|\vec{v}|^2}{r^4\left(\frac{d\theta}{dt}\right)^2} = \frac{\left(\frac{dr}{dt}\right)^2 + \left(r\frac{d\theta}{dt}\right)^2}{r^4\left(\frac{d\theta}{dt}\right)^2} \qquad \left[\text{using } |\vec{v}| = \sqrt{(\dot{r})^2 + (r\dot{\theta})^2}\,\right]$$

$$\Rightarrow \quad \frac{1}{p^2} = \frac{\left(\frac{dr}{dt}\right)^2}{r^4\left(\frac{d\theta}{dt}\right)^2} + \frac{r^2\left(\frac{d\theta}{dt}\right)^2}{r^4\left(\frac{d\theta}{dt}\right)^2}$$

$$\Rightarrow \quad \frac{1}{p^2} = \frac{1}{r^2} + \frac{1}{r^4}\left(\frac{dr}{d\theta}\right)^2 \qquad \text{... (iii)}$$

In view of (i) and (iii) we get,

$$\frac{1}{p^2} = \frac{1}{(r\sin\phi)^2} = \frac{1}{r^2}\operatorname{cosec}^2\phi = \frac{1}{r^2} + \frac{1}{r^4}\left(\frac{dr}{d\theta}\right)^2$$

$$\Rightarrow \quad \frac{1}{r^2}\left(\operatorname{cosec}^2\phi - 1\right) = \frac{1}{r^4}\left(\frac{dr}{d\theta}\right)^2$$

$$\Rightarrow \quad \cot^2\phi = \frac{1}{r^2}\left(\frac{dr}{d\theta}\right)^2$$

$$\Rightarrow \quad \tan\phi = r\frac{d\theta}{dr}. \qquad \text{... (iv)}$$

Corollary. *If* $u = \frac{1}{r}$ *then prove that the pedal equation is* $\frac{1}{p^2} = u^2 + \left(\frac{du}{d\theta}\right)^2$.

Proof. We have $u = \frac{1}{r}$.

$$\Rightarrow \quad \frac{dr}{d\theta} = \frac{d}{d\theta}\left(\frac{1}{u}\right) = -\frac{1}{u^2}\frac{du}{d\theta} \qquad \text{... (i)}$$

$$\frac{1}{p^2}=\frac{1}{r^2}+\frac{1}{r^4}\left(\frac{dr}{d\theta}\right)^2=u^2+u^4\left(-\frac{1}{u^2}\frac{du}{d\theta}\right)^2 \qquad \text{[using (i) and } \frac{1}{r}=u\text{]}$$

$$\Rightarrow \quad \frac{1}{p^2}=u^2+\left(\frac{du}{d\theta}\right)^2$$

Important Results to be used

1. $\frac{1}{p^2}=\frac{1}{r^2}+\frac{1}{r^4}\left(\frac{dr}{d\theta}\right)^2$.

2. $p=r\sin\phi$ and $\tan\phi=r\frac{d\theta}{dr}$

3. $\frac{1}{p^2}=u^2+\left(\frac{du}{d\theta}\right)^2$, where $u=\frac{1}{r}$

REVIEW EXERCISE 2.2

1. Find the pedal equations of the following curve:

(A) Para bola (pole at focus) and $\frac{l}{r}=1+\cos\theta$,

Ans. $p^2=ar$, $\quad l=2a$, is semilatus rectum

(B) Ellipse (pole at focus) and $\frac{l}{r}=1+e\cos\theta$,

Ans. $\frac{b^2}{p^2}=\frac{2a}{r}-1 \quad l$ is semilatus rectum, $e<1$

(C) Hyperbola (pole at focus) and $\frac{l}{r}=1+e\cos\theta$,

Ans. $\frac{b^2}{p^2}=\frac{2a}{r}+1$ for the nearer branch $e>1$

(D) Hyperbola (pole at focus) and $\frac{l}{r}=1+e\cos\theta$,

Ans. $\frac{b^2}{p^2}=1-\frac{2a}{r}$ for the farther branch $e>1$

(E) Ellipse/hyperbola (pole at centre), **Ans.** $\dfrac{a^2b^2}{p^2} = a^2 + b^2 - r^2$

(F) Circle (pole at the circumference $r = 2a\cos\theta$), **Ans.** $r^2 = 2ap$

(G) Equiangular spiral $r = ae^{\theta \cot \alpha}$. **Ans.** $p = r \sin \alpha$

3 Central Orbits

INTRODUCTION

The motions of earth, planets and comets around the sun or the motion of solar system about the centre of mass of milky way galaxy are specific in nature. These motions are governed by specific law where the force causing the motion is always directed towards a fixed point known as Central force. In this chapter we shall study the Central force, motion caused by Central force and path traced due to effect of Central force. Before studying the central force let us revise the *results which are vital in understanding the central orbit:*

Radial Acceleration $(f_r) = \dfrac{d^2r}{dt^2} - r\left(\dfrac{d\theta}{dt}\right)^2 = \ddot{r} - r\dot{\theta}^2$,

Transverse Acceleration $(f_t) = 2\dfrac{dr}{dt}\dfrac{d\theta}{dt} + r\dfrac{d^2\theta}{dt^2} = r\ddot{\theta} + 2\dot{r}\dot{\theta} = \dfrac{1}{r}\dfrac{d\left(r^2\dot{\theta}\right)}{dt}$.

$$\frac{1}{p^2} = \frac{1}{r^2} + \frac{1}{r^4}\left(\frac{dr}{d\theta}\right)^2 = u^2 + \left(\frac{du}{d\theta}\right)^2,$$

where $u = \dfrac{1}{r}$, $p = r\sin\phi$ and $\tan\phi = r\dfrac{d\theta}{dr}$. (Pedal equation)

Radial velocity $= \dfrac{dr}{dt} = \dot{r}$, Transverse velocity $= r\dfrac{d\theta}{dt} = r\dot{\theta}$.

Resultant velocity $\vec{v} = \dot{r}\hat{r} + r\dot{\theta}\hat{s}$, Here $\hat{r}$ and $\hat{s}$ are the unit vectors along and perpendicular to radius vector respectively. $\vec{v}$ will be along the tangent to that point.

§ 1. Central Force

When the motion of a particle is governed by the *force which is always directed towards or away from a fixed point known as Central force* and the fixed point is known as **center of force.** *If the force is always directed towards a fixed point known as* ***Central attraction force.*** If the *force is always directed away from a fixed point known as* ***Central repulsive force.*** For example the motion of the earth around sun is governed by inverse square central attraction force. In figure O is centre of force and f is central attraction force per unit mass (radial acceleration).

Central Orbits

The path described by the particle or object moving in a plane under central force is known as **Central Orbits. In central orbit the transverse O component of acceleration is zero.** For example the motion of the earth around sun is elliptical path and governed by law of central force.

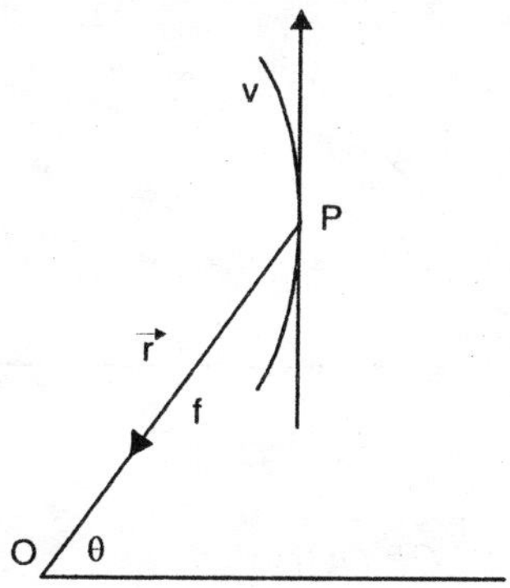

Fig. 3.1

Transverse Acceleration $\left(f_t\right)=\dfrac{1}{r}\dfrac{d\left(r^2\dot{\theta}\right)}{dt}=0$

$\Rightarrow \quad \boldsymbol{r^2\dot{\theta}=h.}$

Where h is a constant.

Theorem. Prove that central orbit is a plane curve.

Proof. In central orbit the *force is always directed towards or away from a fixed point O. Therefore, the moment of the central force about O will be zero.* Let $\vec{r}$ be the position vector of P and particle have resultant velocity $\vec{v}$ at P. We know that

Rate of change of angular momentum = *moment of the force about 0*

$$\Rightarrow \quad \frac{d}{dt}(\vec{r}\times m\vec{v})=0$$

(As the line of action of central force f is passing through O, therefore moment of the force about O is equal to zero)

$\Rightarrow \quad \vec{r} \times m\vec{v} = Constant$

$\Rightarrow \quad \vec{r} \times \vec{v} = Constant.$ *$\vec{r} \times \vec{v}$ (is a vector perpendicular to plane containing $\vec{r}$ and $\vec{v}$ is constant).*

$\Rightarrow \quad \vec{r}$ and $\vec{v}$ *will always be in a plane. Thus* central orbit is a plane curve.

Theorem. *In every central orbit, the radius vector to the centre of force sweeps equal area in equal interval of time* **(Kepler's second law).**

Proof. Let O be the centre of force as origin and affixed line OX as $Q(r+\delta r, \theta+\delta\theta)$ initial line let the particle be at point $P(r,\theta)$ at time t and $P(r,\theta)$ after time $t+\delta t$ particle be at point $Q(r+\delta r, \theta+\delta\theta)$.

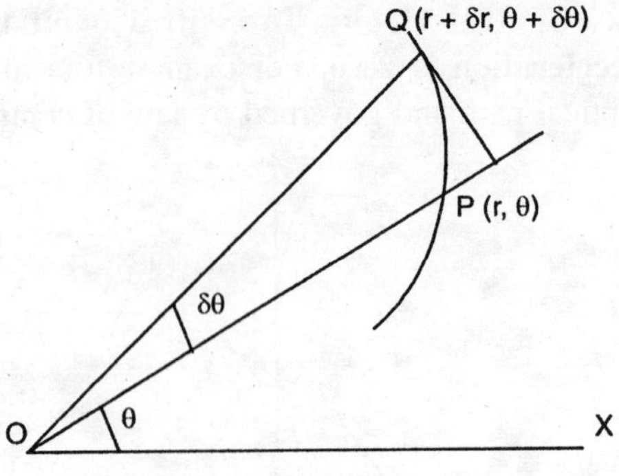

Fig. 3.2

$$\text{Area } \Delta OPQ = \delta A = \frac{1}{2} OP\ OQ \sin\delta\theta$$

$$\Rightarrow \quad \delta A = \frac{1}{2} r(r+\delta r)\sin\delta\theta$$

δA is the area swept during interval δt with the centre of force O by the particle.

Areal velocity = area swept in unit time

$$= lt_{\delta t\to 0}\frac{\delta A}{\delta t} = lt_{\delta t\to 0}\frac{\frac{1}{2}r(r+\delta r)\sin\delta\theta}{\delta t} = lt_{\delta t\to 0}\frac{\frac{1}{2}r(r+\delta r)\delta\theta}{\delta t}$$

$$= \frac{1}{2}\left(r^2\dot{\theta}\right) = \frac{h}{2} \qquad \text{(As } \sin\delta\theta \to \delta\theta \text{ for } \delta\theta \to 0 \text{ and } \delta r\,\delta\theta \to 0)$$

$\Rightarrow$ Areal velocity $= \frac{h}{2}$, h is a constant.

Corollary. *In central orbit the Areal velocity is constant.*

§ 2. A particle moves under the influence of central *attraction* force. Discuss the motion of the particle.

Proof. Let O be the centre of force as origin and fixed line OX as initial line let the particle be at point $P(r, \theta)$ at time t. Let $P(r, \theta)$ be the acceleration (central force per unit mass) of the particle towards O (*in central orbit the transverse O component of acceleration is zero*). Thus for central orbit the

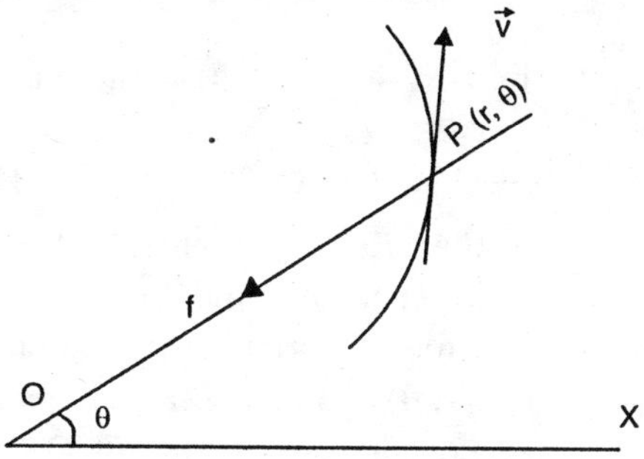

Fig. 3.2

equations of motion are

$$\frac{d^2r}{dt^2} - r\left(\frac{d\theta}{dt}\right)^2 = \ddot{r} - r\dot{\theta}^2 = -f \qquad \text{... (i)}$$

($-f$ indicates that the acceleration is due to *Central attraction force*)

$$r\ddot{\theta} + 2\dot{r}\dot{\theta} = \frac{1}{r}\frac{d\left(r^2\dot{\theta}\right)}{dt} = 0 \qquad \text{... (ii)}$$

(transverse component of acceleration is zero)

On integration (ii) we get,

$$r^2\dot{\theta} = h \qquad \text{... (iii) where } h \text{ is a constant.}$$

$$\dot{\theta} = \frac{h}{r^2} = hu^2 \qquad \text{... (iv) where } u = \frac{1}{r}.$$

$$\dot{r} = \frac{d}{dt}\left(\frac{1}{u}\right) = \frac{d}{d\theta}\left(\frac{1}{u}\right)\frac{d\theta}{dt} = -\dot{\theta}\frac{1}{u^2}\frac{du}{d\theta}$$

$$\Rightarrow \quad \dot{r} = -hu^2\frac{1}{u^2}\frac{du}{d\theta} = -h\frac{du}{d\theta} \qquad \text{... (v) using (iv)}$$

$$\ddot{r} = \frac{d\dot{r}}{dt} = \frac{d}{dt}\left(-h\frac{du}{d\theta}\right) = \frac{d}{d\theta}\left(-h\frac{du}{d\theta}\right)\frac{d\theta}{dt} = -h\frac{d^2u}{d\theta^2}\dot{\theta}$$

$$\Rightarrow \quad \ddot{r} = -h\frac{d^2u}{d\theta^2}hu^2 = -h^2u^2\frac{d^2u}{d\theta^2} \quad \text{... (vi)}$$

In view of (iv) and (vi), (i) reduces into

$$-h^2u^2\frac{d^2u}{d\theta^2} - \frac{1}{u}h^2u^4 = -f$$

$$\Rightarrow \quad f = h^2u^2\left(\frac{d^2u}{d\theta^2} + u\right) \qquad \text{... (vii) Required equation of motion.}$$

Note. *If f is a Central repulsive force replace f by –f in (vii).*

§ 3. A particle moves under the influence of central attraction force. Discuss the motion of the particle in terms of Pedal equation.

Proof. Let O be the centre of force as origin and fixed line OX as initial line let the particle be at point $P(r, \theta)$ at time *t*. Let $P(r, \theta)$ be the acceleration (central force per unit mass) of the particle towards O. Let $OM = p$ be the length of perpendicular from O upon the tangent at P. Then by Pedal equation we have

$$\frac{1}{p^2} = u^2 + \left(\frac{du}{d\theta}\right)^2 \qquad \text{... (i), where } u = \frac{1}{r}$$

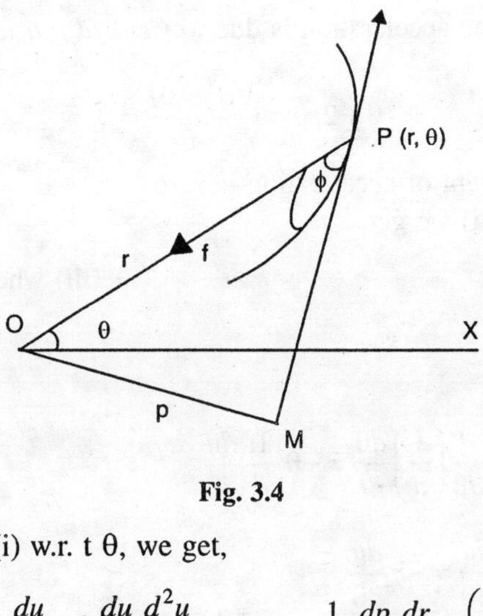

Fig. 3.4

Differentiating (i) w.r. t θ, we get,

$$-\frac{2}{p^3}\frac{dp}{d\theta} = 2u\frac{du}{d\theta} + 2\frac{du}{d\theta}\frac{d^2u}{d\theta^2} \quad \Rightarrow \quad -\frac{1}{p^3}\frac{dp}{dr}\frac{dr}{d\theta} = \left(u + \frac{d^2u}{d\theta^2}\right)\frac{du}{d\theta}$$

$$\Rightarrow \quad -\frac{1}{p^3}\frac{dp}{dr}\left(\frac{-1}{u^2}\right)\frac{du}{d\theta}=\left(u+\frac{d^2u}{d\theta^2}\right)\frac{du}{d\theta},$$

$$\left(\text{using}\ \frac{du}{d\theta}=\frac{d}{d\theta}(1/r)=\frac{-1}{r^2}\frac{dr}{d\theta}=-u^2\frac{dr}{d\theta}\right)$$

$$\Rightarrow \quad \frac{1}{p^3}\frac{dp}{dr}=u^2\left(u+\frac{d^2u}{d\theta^2}\right) \qquad \text{... (ii)}$$

We know that $=h^2u^2\left(\frac{d^2u}{d\theta^2}+u\right)$, Putting f *in (ii) we get,* $\frac{1}{p^3}\frac{dp}{dr}=\frac{f}{h^2}$

$$\Rightarrow \quad f=\frac{h^2}{p^3}\frac{dp}{dr}$$ ... (iii) Required equation of motion in Pedal form.

§ 4. A particle moves under the influence of central *attraction* force. Derive the relation between the resultant velocity *v* of particle in terms of Pedal length *p*.

Proof. Let O be the centre of force as origin and fixed line OX as initial line let the particle be at point $P(r, \theta)$ at time t. Let f be the acceleration (central force per unit mass) of the particle towards O. Let $OM = p$ be the length of perpendicular from O upon the tangent at P and let radial and transverse velocities be $\dot{r}$ and $r\dot{\theta}$ respectively.

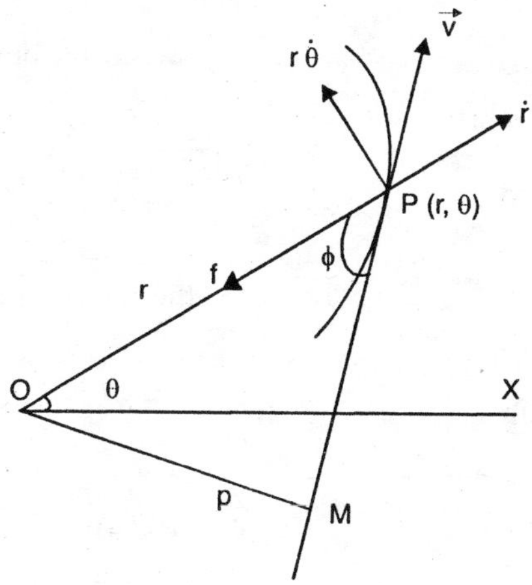

Fig. 3.5

Taking moment of the velocities about O we get

$v \times p = r \times r\dot{\theta}$

$\Rightarrow \quad vp = r^2\dot{\theta} = h$ (using the condition for central orbit $r^2\dot{\theta} = h$ = constant)

$\Rightarrow \quad v = \dfrac{h}{p}$ ... (i) Required relation between v and p.

$$v^2 = \frac{h^2}{p^2} = h^2\left(u^2 + \left(\frac{du}{d\theta}\right)^2\right) \quad \text{... (ii)}$$

Corollary. *In central orbit the velocity at any point varies inversely as the length of perpendicular from centre to the tangent at that point.*

Some important curves and their polar/pedal equations

Curve	*polar equation/pedal equation*
1. Parabola (pole at focus) and	$\frac{l}{r} = 1 + \cos\theta, \quad p^2 = ar$ $l = 2a$, is semilatus rectum
2. Ellipse (pole at focus) and	$\frac{l}{r} = 1 + e\cos\theta, \quad \frac{b^2}{p^2} = \frac{2a}{r} - 1$ l is semilatus rectum, $e < 1$
3. Hyperbola (pole at focus) and $\frac{l}{r} = 1 + e\cos\theta,$	$\frac{b^2}{p^2} = \frac{2a}{r} + 1$ for the nearer branch, $e > 1$
4. Hyperbola (pole at focus) and $\frac{l}{r} = 1 + e\cos\theta,$	$\frac{b^2}{p^2} = 1 - \frac{2a}{r}$ for the farther branch, $e > 1$
5. Ellipse/hyperbola (pole at centre)	$\frac{a^2b^2}{p^2} = a^2 + b^2 - r^2$
6. Circle (pole at the circumference)	$r = 2a\cos\theta, \quad r^2 = 2ap$
7. Equiangular spiral	$r = ae^{\theta\cot\alpha}, \quad p = r\sin\alpha$

Note. *Pole implies the centre of force.*

Important results to be used

1. $r^2\dot{\theta} = h$. Where h is a constant, Areal velocity $= \frac{h}{2}$.

2. $f = h^2u^2\left(\frac{d^2u}{d\theta^2} + u\right)$, $\qquad f = \frac{h^2}{p^3}\frac{dp}{dr}$

3. $v^2 = \frac{h^2}{p^2} = h^2\left(u^2 + \left(\frac{du}{d\theta}\right)^2\right)$, $p = r\sin\phi$ and $\tan\phi = r\frac{d\theta}{dr}$

Here $u = 1/r$.

Type I. To find the force, if the orbit is given in polar form. The law of force can be obtained by either of the two methods as illustrated in the examples.

ILLUSTRATIVE EXAMPLES

Example 1 A particle describes the equiangular spiral $r = ae^{\theta\cot\alpha}$ under a central force f to the pole. Find the law of force.

Solution We have $r = ae^{\theta\cot\alpha}$

$$\Rightarrow \quad u = \frac{1}{r} = \frac{e^{-\theta\cot\alpha}}{a} \qquad \dots (1)$$

Differentiating twice (1) w. r. t. θ,we get, $\frac{du}{d\theta} = \frac{-\cot\alpha}{a}e^{-\theta\cot\alpha} = -u\cot\alpha$

$$\Rightarrow \quad \frac{d^2u}{d\theta^2} = -\frac{du}{d\theta}\cot\alpha = u\cot^2\alpha \qquad \dots (2)$$

Law of force of central orbit is given by

$$f = h^2u^2\left(\frac{d^2u}{d\theta^2} + u\right) \qquad \dots (3)$$

In view of (1) and (2), we get $f = h^2u^2(u\cot^2\alpha + u) = h^2u^3(\cot^2\alpha + 1)$

$$\Rightarrow \quad f = h^2u^3\operatorname{cosec}^2\alpha$$

$$\Rightarrow \quad f = \frac{h^2\operatorname{cosec}^2\alpha}{r^3}$$

$\Rightarrow \quad f = \dfrac{\mu}{r^3}$ ... (4), where $\mu = h^2 \operatorname{cosec}^2 \alpha$, h and α are constants.

Thus in view of (4) we can say that the central force varies inversely cubic power of the distance from the pole.

Alternate method

We know that the pedal equation equiangular spiral $r = ae^{\theta \cot \alpha}$ is

$$p = r \sin \alpha \quad \text{... (1)}$$

Differentiating (1) w.r. t. r, we get, $\dfrac{dp}{dr} = \sin \alpha$ ... (2)

Law of force of central orbit is given by

$$f = \frac{h^2}{p^3} \frac{dp}{dr}$$

In view of (1) and (2) we get, $f = \dfrac{h^2}{r^3 \sin^3 \alpha} \sin \alpha$

$\Rightarrow \quad f = \dfrac{h^2}{r^3} \operatorname{cosec}^2 \alpha \quad \Rightarrow \quad f = \dfrac{\mu}{r^3}$ ... (3)

where $\mu = h^2 \operatorname{cosec}^2 \alpha h$ and α are constants.

Example 2 A particle describes the circle with the pole on the circumference $r = 2a \cos \theta$ under a central force f to the pole. Find the law of force.

Solution We have $r = 2a \cos \theta$

$$\Rightarrow \quad u = \frac{1}{r} = \frac{\sec \theta}{2a} \quad \text{... (1)}$$

Differentiating twice (1) w.r. t. θ, we get,

$$\frac{du}{d\theta} = \frac{\sec \theta \tan \theta}{2a} = u \tan \theta$$

$$\Rightarrow \quad \frac{d^2u}{d\theta^2} = \frac{du}{d\theta} \tan \theta + u \sec^2 \theta = u(\tan^2 \theta + \sec^2 \theta) \quad \text{... (2)}$$

Law of force of central orbit is given by $f = h^2 u^2 \left(\dfrac{d^2u}{d\theta^2} + u \right)$ ... (3)

In view of (1) and (2), we get

$$f = h^2u^2\left(u\left(\tan^2\theta + \sec^2\theta\right) + u\right) = h^2u^3\left(2\sec^2\theta\right) = 2h^2u^3 4a^2u^2$$

$$\Rightarrow \quad f = 8a^2h^2u^5 \quad \Rightarrow \quad f = \frac{8a^2h^2}{r^5} \quad \Rightarrow \quad f = \frac{\mu}{r^5} \qquad \dots (4),$$

where $\mu = 8a^2h^2$, h and a are constants.

Thus in view of (4) we can say that the central force varies inversely fifth power of the distance from the pole.

Alternate method

We know that the pedal equation of circle with the pole on the circumference $r = 2a\cos\theta$ is $r^2 = 2ap$

$$\Rightarrow \qquad p = \frac{r^2}{2a} \qquad \dots (1)$$

Differentiating (1) w.r. t. *r*, we get,

$$\frac{dp}{dr} = \frac{r}{a} \qquad \dots (2)$$

Law of force of central orbit is given by $f = \frac{h^2}{p^3}\frac{dp}{dr}$

In view of (1) and (2) we get, $f = \frac{8a^3h^2}{r^6}\frac{r}{a}$

$$\Rightarrow \qquad f = \frac{8a^2h^2}{r^5}$$

$$\Rightarrow \qquad f = \frac{\mu}{r^5} \qquad \dots (3)$$

where $\mu = 8a^2h^2$, h and a are constants.

Example 3. A particle describes the ellipse with the pole on the focus $\frac{l}{r} = 1 + e\cos\theta$ under a central force *f* to the pole. Find the law of force.

Solution We have $\frac{l}{r} = 1 + e\cos\theta$

$$\Rightarrow \qquad u = \frac{1}{r} = \frac{1}{l} + \frac{e\cos\theta}{l} \qquad \dots (1)$$

Differentiating twice (1) w.r. t θ, we get,

$$\frac{du}{d\theta} = -\frac{e\sin\theta}{l}$$

$$\Rightarrow \qquad \frac{d^2u}{d\theta^2} = -\frac{e\cos\theta}{l} \qquad \dots (2)$$

Law of force of central orbit is given by

$$f = h^2u^2\left(\frac{d^2u}{d\theta^2} + u\right) \qquad \dots (3)$$

In view of (1) and (2), we get

$$f = h^2u^2\left(\frac{-e\cos\theta}{l} + \frac{1}{l} + \frac{e\cos\theta}{l}\right) = \frac{h^2u^2}{l}$$

$$\Rightarrow \quad f = \frac{h^2u^2}{l} = \frac{h^2}{l}\frac{1}{r^2}$$

$$\Rightarrow \quad f = \frac{\mu}{r^2} \qquad \dots (4), \text{ where } \mu = \frac{h^2}{l},\ h \text{ and } l \text{ are constants.}$$

Thus in view of (4) we can say that the central force varies inverse square of the distance from the pole.

Alternate method

We know that the pedal equation of the ellipse with the pole at the focus $\frac{l}{r} = 1 + e\cos\theta$, is

$$\frac{b^2}{p^2} = \frac{2a}{r} - 1 \qquad \dots (1)$$

Differentiating (1) w.r. t. r, we get,

$$\frac{-2b^2}{p^3}\frac{dp}{dr} = -\frac{2a}{r^2} \Rightarrow \frac{1}{p^3}\frac{dp}{dr} = \frac{a}{b^2r^2} \qquad \dots (2)$$

Law of force of central orbit is given by

$$f = \frac{h^2}{p^3}\frac{dp}{dr}$$

In view of (2) we get, $f = \dfrac{h^2 a}{b^2 r^2}$

$$\Rightarrow \quad f = \frac{h^2}{l}\frac{1}{r^2}, \text{ where } l = \frac{b^2}{a} = \text{semi latus rectum of ellipse.}$$

$$\Rightarrow \quad f = \frac{\mu}{r^2} \qquad \ldots (3)$$

where $\mu = \dfrac{h^2}{l}$, h and l are constants.

Example 4 A particle describes the curve $r^n = a^n \cos n\theta$ under a central force f to the pole. Find the law of force.

Solution We have $r^n = a^n \cos n\theta \Rightarrow u^n a^n \cos n\theta = 1 \Rightarrow \sec n\theta = u^n a^n$ ... (1)

Taking logarithm

$$n \log u + n \log a + \log \cos n\theta = 0 \qquad \ldots (2)$$

Differentiating twice (2) w.r. t. θ, we get, $\dfrac{n}{u}\dfrac{du}{d\theta} = n\dfrac{\sin n\theta}{\cos n\theta}$

$$\Rightarrow \quad \frac{du}{d\theta} = u \tan n\theta$$

$$\Rightarrow \quad \frac{d^2u}{d\theta^2} = nu\sec^2 n\theta + \frac{du}{d\theta}\tan n\theta = u(n\sec^2 n\theta + \tan^2 n\theta) \qquad \ldots (3)$$

Law of force of central orbit is given by

$$f = h^2u^2\left(\frac{d^2u}{d\theta^2} + u\right) \qquad \ldots (4)$$

In view of (1) and (3), we get

$$f = h^2u^2\left\{u(n\sec^2 n\theta + \tan^2 n\theta) + u\right\} = h^2u^3(n+1)\sec^2 n\theta$$

$$\Rightarrow \quad f = h^2u^3(n+1)u^{2n}a^{2n} = (n+1)a^{2n}h^2u^{2n+3}$$

$$\Rightarrow \quad f = \frac{(n+1)a^{2n}h^2}{r^{2n+3}}$$

$\Rightarrow f = \dfrac{\mu}{r^{2n+3}}$ ... (4), where $\mu = (n+1)a^{2n}h^2$, h, n and a are constants.

Thus in view of (4) we can say that the central force varies inverse square of the distance from the pole.

Alternate method

We have $$r^n = a^n \cos n\theta \quad \ldots (1)$$

Taking logarithm

$$n \log r = n \log a + \log \cos n\theta \quad \ldots (2)$$

Differentiating (2) w.r. t. θ, we get, $\dfrac{n}{r}\dfrac{dr}{d\theta} = -n\dfrac{\sin n\theta}{\cos n\theta}$

$$\Rightarrow \quad r\frac{d\theta}{dr} = -\cot n\theta \tan\phi = r\frac{d\theta}{dr} = -\cot n\theta = \tan\left(\frac{\pi}{2} + n\theta\right)$$

$$\Rightarrow \quad \phi = \left(\frac{\pi}{2} + n\theta\right) \quad \ldots (3)$$

Hence in view of (1) and (3) the pedal equation of the curve is

$$p = r\sin\phi = r\sin\left(\frac{\pi}{2} + n\theta\right) = r\cos n\theta\, p = \frac{r^{n+1}}{a^n} \quad \ldots (4)$$

Differentiating (4) w.r. t. r, we get,

$$\frac{dp}{dr} = (n+1)\frac{r^n}{a^n} \quad \ldots (5)$$

Law of force of central orbit is given by $f = \dfrac{h^2}{p^3}\dfrac{dp}{dr}$

In view of (4) and (5) we get,

$$\Rightarrow \quad f = \frac{(n+1)a^{2n}h^2}{r^{2n+3}}$$

$$\Rightarrow \quad f = \frac{\mu}{r^{2n+3}} \quad \ldots (6)$$

where $\mu = (n+1)a^{2n}h^2$, h, n and a are constants. .

Type II. To find the Orbit, if the law of central force is given in polar form.

Algorithm to find the central orbit

Step 1. Write the law of force $\left(\dfrac{d^2u}{d\theta^2} + u\right) = \dfrac{f}{h^2u^2}$ ... (1), where $h = vp$.

Step 2. Multiplying both sides of (1) by $2\frac{du}{d\theta}$ and then integrate the result w.r. t. θ.

Step 3. The resulting differential equation will be

$\left(u^2+\left(\frac{du}{d\theta}\right)^2\right)=2\int\frac{f}{h^2u^2}du$. Which on integration yields $u=F(\theta)$.

Step 4. Replace u by $\frac{1}{r}$ in $u=F(\theta)$.

Note. *If RHS of (1) contains constant only or the term containing first degree of u then it is advisable to solve the second order differential equation directly. Following illustrations explain the use of differential equation directly.*

ILLUSTRATIVE EXAMPLES

Example 1 Find the orbit, if the central force varies inversely as the square of the distance from a fixed point.

Solution I[st] method

we have $f=\frac{\mu}{r^2}=\mu u^2$

$$\Rightarrow \quad \left(\frac{d^2u}{d\theta^2}+u\right)=\frac{f}{h^2u^2}=\frac{\mu u^2}{h^2u^2} \quad \Rightarrow \quad \left(\frac{d^2u}{d\theta^2}+u\right)=\frac{\mu}{h^2} \qquad \ldots (1)$$

Multiplying both sides of (1) by $2\frac{du}{d\theta}$ and then integrate the result w.r. t. θ

$$2\frac{du}{d\theta}\frac{d^2u}{d\theta^2}+2\frac{du}{d\theta}u=\frac{\mu}{h^2}2\frac{du}{d\theta}$$

$$\Rightarrow \quad \left(\frac{du}{d\theta}\right)^2+u^2=2\frac{\mu}{h^2}u$$

$$\Rightarrow \quad \left(\frac{du}{d\theta}\right)^2=-\left(u^2-2\frac{\mu}{h^2}u\right)=\frac{\mu^2}{h^4}-\left(u-\frac{\mu}{h^2}\right)^2$$

$$\Rightarrow \quad \int\frac{du}{\sqrt{\frac{\mu^2}{h^4}-\left(u-\frac{\mu}{h^2}\right)^2}}=\pm\int d\theta+C \Rightarrow \cos^{-1}\frac{\left(u-\frac{\mu}{h^2}\right)}{\frac{\mu}{h^2}}=\mp\theta+C$$

$$\Rightarrow \quad \left(u - \frac{\mu}{h^2}\right) = \frac{\mu}{h^2} A\cos(\theta - \alpha)$$

$$\Rightarrow \quad \frac{h^2/\mu}{r} - 1 = A\cos(\theta - \alpha) \qquad \text{(Multiplying by } h^2/\mu \text{ and putting } u = \frac{1}{r}\text{)}$$

$$\Rightarrow \quad \frac{h^2/\mu}{r} = 1 + A\cos(\theta - \alpha)$$

$$\Rightarrow \quad \frac{l}{r} = 1 + A\cos(\theta - \alpha) \text{ where } l = h^2/\mu.$$

Thus the orbit is a conic, focus being the pole.

IInd method

We have $f = \frac{\mu}{r^2} = \mu u^2$

$$\Rightarrow \quad \left(\frac{d^2u}{d\theta^2} + u\right) = \frac{f}{h^2u^2} = \frac{\mu u^2}{h^2u^2}$$

$$\Rightarrow \quad \left(\frac{d^2u}{d\theta^2} + u\right) = \frac{\mu}{h^2} \text{ ... (1) (see note RHS is constant)}$$

$$\Rightarrow \quad \frac{d^2}{d\theta^2}\left(u - \frac{\mu}{h^2}\right) + \left(u - \frac{\mu}{h^2}\right) = 0, \frac{d^2}{d\theta^2}\left(\frac{\mu}{h^2}\right) = 0 \text{ as } \frac{\mu}{h^2} \text{ is a constant)}$$

$$\Rightarrow \quad \frac{d^2}{d\theta^2}\left(u - \frac{\mu}{h^2}\right) = -\left(u - \frac{\mu}{h^2}\right) \qquad \text{... (2)}$$

Let $\left(u - \frac{\mu}{h^2}\right) = z$, then (2) reduces to $\frac{d^2z}{d\theta^2} = z$

$$\Rightarrow \quad z = B\cos\theta + C\sin\theta = A\cos(\theta - \alpha)$$

$$\Rightarrow \quad \left(u - \frac{\mu}{h^2}\right) = A\cos(\theta - \alpha)$$

$$\Rightarrow \quad \frac{h^2/\mu}{r} - 1 = A_1\cos(\theta - \alpha)$$

(Multiplying by h^2/μ and putting $u = \frac{1}{r}$ and $A_1 = Ah^2/\mu$)

$$\Rightarrow \quad \frac{h^2/\mu}{r} = 1 + A_1 \cos(\theta - \alpha)$$

$$\Rightarrow \quad \frac{l}{r} = 1 + A_1 \cos(\theta - \alpha) \text{ where } l = h^2/\mu .$$

Thus the orbit is a conic, focus being the pole.

Example 2 Find the orbit, if the central force varies inversely as the cube of the distance from a fixed point.

Solution We have $f = \frac{\mu}{r^3} = \mu u^3$

$$\Rightarrow \quad \left(\frac{d^2u}{d\theta^2} + u\right) = \frac{f}{h^2u^2} = \frac{\mu u^3}{h^2u^2}$$

$$\Rightarrow \quad \left(\frac{d^2u}{d\theta^2} + u\right) = \frac{\mu}{h^2}u$$ (see note RHS is first degree of u)

$$\Rightarrow \quad \frac{d^2u}{d\theta^2} = -\left(1 - \frac{\mu}{h^2}\right)u \qquad \ldots (1)$$

The solution of (1) depends on the value of $\frac{\mu}{h^2}$. There are three possibilities:

Case 1. $\frac{\mu}{h^2} > 1$ i.e. $\frac{\mu}{h^2} - 1 > 0$, (1) reduces to

$$\frac{d^2u}{d\theta^2} = n^2u$$ ... (2), where $\frac{\mu}{h^2} - 1 = n^2$. Its solution is

$$u = c_1e^{n\theta} + c_2e^{-n\theta} = A\cosh n\theta + B\sinh n\theta$$

$$\frac{1}{r} = A\cosh n\theta + B\sinh n\theta.$$

Case 2. $\frac{\mu}{h^2} < 1$ i.e., $\frac{\mu}{h^2} - 1 < 0$, (1) reduces to $\frac{d^2u}{d\theta^2} = -n^2u \qquad \ldots (3)$

where $\frac{\mu}{h^2}-1=-n^2$. Its solution is $u=A\cos n\theta+B\sin n\theta$

$$\frac{1}{r}=A\cos n\theta+B\sin n\theta.$$

Case 3. $\frac{\mu}{h^2}=1$ i.e., $\frac{\mu}{h^2}-1=0$, (1) reduces to

$$\frac{d^2u}{d\theta^2}=0 \quad \text{... (4). Its solution is } u=A\theta+B$$

$$\frac{1}{r}=A\theta+B.$$

Note. *The nature and the shape of the orbit can also be determined from solving the differential equation* $f=\frac{h^2}{p^3}\frac{dp}{dr}$ *for p and r comparing it from the pedal equation in given table.*

Example 3 Find the orbit, if the central force varies as the distance from a fixed point.

Solution We have $f=\frac{h^2}{p^3}\frac{dp}{dr}=\mu r \Rightarrow \frac{h^2}{p^3}dp=\mu r\,dr$

On integration we get

$$\int\frac{h^2}{p^3}dp=\int\mu r dr$$

$$\Rightarrow \quad -\frac{h^2}{2p^2}=\frac{\mu r^2}{2}+C$$

$$\Rightarrow \quad \frac{h^2}{\mu p^2}=-r^2+C' \quad \text{... (1)}$$

Equation (1) is analogous to pedal equation of an ellipse, pole at centre, if $\frac{h^2}{\mu}=a^2b^2$ and $c'=a^2+b^2$.

Thus the orbit is an ellipse where the centre of force is at the centre of ellipse.

Alternative method for Example 1

Solution We have $f = \frac{h^2}{p^3}\frac{dp}{dr} = \frac{\mu}{r^2}$

$$\Rightarrow \quad \frac{h^2}{p^3}dp = \frac{\mu}{r^2}dr$$

On integration we get

$$\int \frac{h^2}{p^3}dp = \int \frac{\mu}{r^2}dr$$

$$\Rightarrow \quad -\frac{h^2}{2p^2} = -\frac{\mu}{r} + c$$

$$\Rightarrow \quad \frac{h^2}{p^2} = \frac{\mu}{r} + c' \qquad \dots (1)$$

Case 1. Equation (1) is analogous to pedal equation of an ellipse

$$\frac{b^2}{p^2} = \frac{2a}{r} - 1 \text{ if } c' < 0.$$

Case 2. Equation (1) is analogous to pedal equation of a hyperbola

$$\frac{b^2}{p^2} = \frac{2a}{r} + 1 \text{ if } c' > 0.$$

Case 3. Equation (1) is analogous to pedal equation of a parabola

$$p^2 = ar \text{ if } c' = 0.$$

Alternative method for Example 2 Left for Readers.

Type III. To find the Orbit, if the law of central force and velocity at any point are given.

Algorithm to find the central orbit.

Step 1. Find h^2 using $h^2 = p^2v^2 =$ where $p = r\sin\phi$ and v is velocity at the given point.

Step 2. Put the value of f in $f = \frac{h^2}{p^3}\frac{dp}{dr}$ and integrate the result.

Step 3. Substitute the value of h^2 in the result obtained in step 1 and apply the condition $r = a$, $\theta = 0$, hence evaluate the integration constant.

Step 4. Use $\frac{1}{p^2} = \frac{1}{r^2} + \frac{1}{r^4}\left(\frac{dr}{d\theta}\right)^2$ and find $\frac{dr}{d\theta}$.

Step 5. Integrate the result of step-4 and apply *apse* condition $p = r = a$ at $\theta = 0$, hence evaluate the integration constant.

ILLUSTRATIVE EXAMPLES

Example 1 A particle acted on by a central attraction force μ/r^3 is projected with a velocity $\sqrt{\mu/a}$ at a angle of $\pi/4$ with its initial distance a from the centre of force. Show that the path is the equiangular spiral.

Solution Let $OP = a$, the length of the perpendicular from O on the tangent at $P = p_0$ and O is taken as pole.

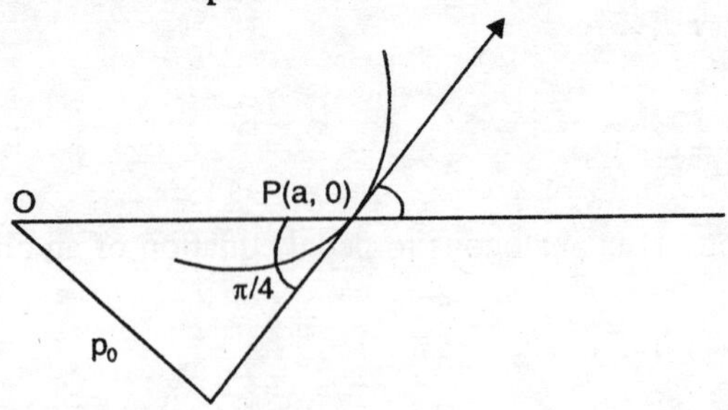

Fig. 3.6

The angle between the tangent at P and the radius vector $OP = \phi = \frac{\pi}{4}$ and velocity at $P = v = \sqrt{\mu/a}$.

Using $h^2 = p^2v^2$ where $p = r\sin\phi$

$$p_0 = r\sin\phi = a\sin\frac{\pi}{4} = \frac{a}{\sqrt{2}} \qquad \text{... (1)}$$

$(p = p_0,\ r = a \text{ and } \phi = \pi/4)$

$$\sqrt{\mu}/ah^2 = p_0^2v^2 = \frac{a^2}{2}\left(\sqrt{\mu}/a\right)^2 = \frac{\mu}{2} \qquad \text{... (2)}$$

We have $f = \frac{h^2}{p^3}\frac{dp}{dr} = \frac{\mu}{r^3} \quad \Rightarrow \quad \frac{h^2}{p^3}dp = \frac{\mu}{r^3}dr$

On integration we get $\int \frac{h^2}{p^3}dp = \int \frac{\mu}{r^3}dr$

$$\Rightarrow \quad -\frac{h^2}{2p^2} = -\frac{\mu}{2r^2} + c \quad \Rightarrow \quad \frac{h^2}{p^2} = \frac{\mu}{r^2} + c'$$

Using (2) we get,

$$\Rightarrow \quad \frac{\mu}{2p^2} = \frac{\mu}{r^2} + c' \qquad \ldots (3)$$

We have from (1) $p = \frac{a}{\sqrt{2}}$ when $r = a$. Using these conditions in (3) we get,

$$\Rightarrow \quad \frac{\mu}{2\left(\frac{a}{\sqrt{2}}\right)^2} = \frac{\mu}{a^2} + c' \quad \Rightarrow \quad c' = 0 \qquad \ldots (4)$$

In view of (3) and (4) we get,

$$\Rightarrow \quad \frac{\mu}{2}\left(\frac{1}{r^2} + \frac{1}{r^4}\left(\frac{dr}{d\theta}\right)^2\right) = \frac{\mu}{r^2}$$

$$\Rightarrow \quad \frac{1}{r^4}\left(\frac{dr}{d\theta}\right)^2 = \frac{2}{r^2} - \frac{1}{r^2} \Rightarrow \left(\frac{dr}{d\theta}\right)^2 = r^2$$

$$\Rightarrow \quad \frac{dr}{d\theta} = \pm r$$

Taking negative sign only as the rate of change of r w.r. t θ is negative.

$$\int \frac{dr}{r} = -\int d\theta + A$$

$$\Rightarrow \quad \log r = -\theta + A \qquad \ldots (5)$$

Initially $\theta = 0, r = a$ we get $A = \log a$

$$\Rightarrow \quad \log r = -\theta + \log a$$

$$\Rightarrow \quad \log \frac{r}{a} = -\theta$$

$$\Rightarrow \quad r = ae^{-\theta}.$$

Which is the equation of an equiangular spiral.

Example 2. A particle acted on by a central attraction force $\mu\left(\frac{3}{r^3}+\frac{2a^2}{r^5}\right)$ is projected with a velocity $5\sqrt{\mu/a}$ at a angle of $\cot^{-1} 2$ with its initial distance a from the centre of force. Find the path of the particle.

Solution Let $OP = a$, the length of the perpendicular from O on the tangent at $P = p_0$ and O is taken as pole.

The angle between the tangent at P and the radius vector $OP = \phi = \cot^{-1} 2$ and velocity at $P = v = 5\sqrt{\mu/a}$.

Using $h^2 = p^2v^2$ where $p = r\sin\phi$,

$$\cot\phi = 2 \Rightarrow 1+\cot^2\phi = \operatorname{cosec}^2\phi = 1+4=5$$

$$\Rightarrow \quad \sin^2\phi = \frac{1}{5} \Rightarrow \sin\phi = \frac{1}{\sqrt{5}} \qquad \ldots (1)$$

$$p_0 = r\sin\phi = a\frac{1}{\sqrt{5}} = \frac{a}{\sqrt{5}} \qquad \ldots (2)$$

$$\left(p = p_0,\ r = a \text{ and } \phi = \cot^{-1} 2\right)$$

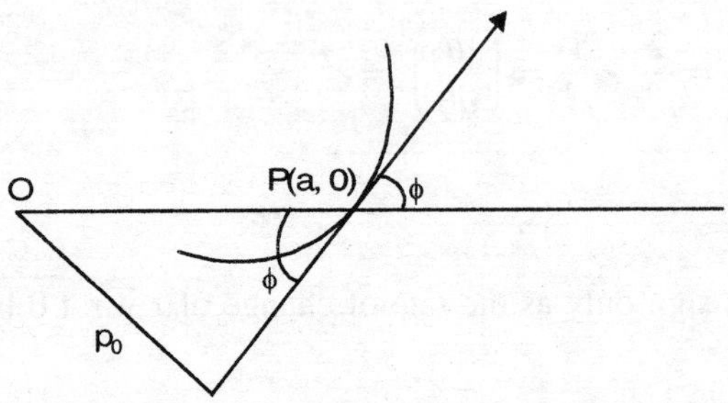

Fig. 3.7

$$5\sqrt{\mu/ah^2} = p_0^2v^2 = \frac{a^2}{5}\left(5\sqrt{\mu/a}\right)^2 = \mu \qquad \ldots (3)$$

We have $f = \dfrac{h^2}{p^3}\dfrac{dp}{dr} = \mu\left(\dfrac{3}{r^3}+\dfrac{2a^2}{r^5}\right)$

$$\Rightarrow \quad \frac{h^2}{p^3}dp = \mu\left(\frac{3}{r^3}+\frac{2a^2}{r^5}\right)dr$$

On integration we get

$$\int \frac{h^2}{p^3} dp = \int \mu \left(\frac{3}{r^3} + \frac{2a^2}{r^5} \right) dr$$

$$\Rightarrow \quad -\frac{h^2}{2p^2} = -\mu \left(\frac{3}{2r^2} + \frac{2a^2}{4r^4} \right) + c$$

$$\Rightarrow \quad \frac{h^2}{p^2} = \mu \left(\frac{3}{r^2} + \frac{a^2}{r^4} \right) + c'$$

Using (3) we get,

$$\Rightarrow \quad \frac{\mu}{p^2} = \mu \left(\frac{3}{r^2} + \frac{a^2}{r^4} \right) + c' \qquad \text{... (4)}$$

We have from (2) $p = \dfrac{a}{\sqrt{5}}$ when $r = a$. Using these conditions in (3) we get,

$$\Rightarrow \quad \frac{\mu}{\left(\frac{a}{\sqrt{5}} \right)^2} = \frac{4\mu}{a^2} + c' \Rightarrow \frac{5\mu}{a^2} = \frac{4\mu}{a^2} + c'$$

$$\Rightarrow \quad c' = \frac{\mu}{a^2} \qquad \text{... (5)}$$

In view of (4) and (5) we get,

$$\Rightarrow \quad \mu \left(\frac{1}{r^2} + \frac{1}{r^4} \left(\frac{dr}{d\theta} \right)^2 \right) = \mu \left(\frac{3}{r^2} + \frac{a^2}{r^4} \right) + \frac{\mu}{a^2}$$

$$\Rightarrow \quad \frac{1}{r^4} \left(\frac{dr}{d\theta} \right)^2 = \frac{2}{r^2} + \frac{a^2}{r^4} + \frac{1}{a^2}$$

$$\Rightarrow \quad \left(\frac{dr}{d\theta} \right)^2 = \frac{2r^2a^2 + a^4 + r^4}{a^2} = \frac{(a^2 + r^2)^2}{a^2}$$

$$\Rightarrow \quad \frac{dr}{d\theta} = \frac{(a^2 + r^2)}{a}$$

$$\int \frac{dr}{(a^2+r^2)} = \frac{1}{a}\int d\theta + A$$

$$\Rightarrow \quad \frac{1}{a}\tan^{-1}\left(\frac{r}{a}\right) = \frac{1}{a}\theta + A \qquad \ldots (6)$$

Initially $\theta = 0$, $r = a$ we get $A = \dfrac{\pi}{4a}$

$$\Rightarrow \quad \frac{1}{a}\tan^{-1}\left(\frac{r}{a}\right) = \frac{1}{a}\theta + \frac{\pi}{4a}$$

$$\Rightarrow \quad \frac{r}{a} = \tan\left(\theta + \frac{\pi}{4}\right)$$

$\Rightarrow r = a\tan\left(\theta + \dfrac{\pi}{4}\right)$. Which is the required equation of the path.

REVIEW EXERCISE 3.1

1. Define central force. Prove that central orbit is a plane curve.
2. Show that in central orbit the Areal velocity is constant.
3. A particle moves under the influence of central *attraction* force. Discuss the motion of the particle.
4. Show that in central orbit the velocity at any point varies inversely as the length of perpendicular from centre to the tangent at that point.
5. A particle describes the curve $r^2 = a^2\cos 2\theta$ under a central force f to the pole. Find the law of force. **(Ans.** $f = \dfrac{\mu}{r^7}$)
6. A particle describes the cardioid $r = a(1+\cos\theta)$ under a central force f to the pole. Find the law of force. **(Ans.** $f = \dfrac{\mu}{r^4}$)
7. A particle describes the parabola with the pole on the focus $\dfrac{l}{r} = 1 + \cos\theta$ under a central force f to the pole. Find the law of force. **(Ans.** $f = \dfrac{\mu}{r^2}$)
8. A particle describes the curve $r = a\cos n\theta$ under a central force f to the pole. Show that central force f is directly proportional to $\dfrac{2n^2a^2}{r^5} - \dfrac{n^2-1}{r^3}$.

9. A particle describes the curve $r^n \cos n\theta = a^n$ under a central force f to the pole. Show that central force f is directly proportional to r^{2n-3}.
10. The pedal equation of an ellipse with pole at the centre is

$\frac{a^2b^2}{p^2} = a^2 + b^2 - r^2$. Find the law of force. (**Ans.** $f = \mu r$.)

11. The pedal equation of a parabola with pole at the focus is $p^2 = ar$. Find the law of force. (**Ans.** $f = \frac{\mu}{r^2}$)

12. The pedal equation of an ellipse with pole at the focus is $\frac{b^2}{p^2} = \frac{2a}{r} - 1$.

Find the law of force. (**Ans.** $f = \frac{\mu}{r^2}$)

13. Find the orbit, if the central force varies inversely as the cube of the distance from a fixed point.
14. Find the orbit, if the central force varies inversely as the square of the distance from a fixed point.
15. Find the orbit, if the central force varies as the distance from a fixed point. (**Ans.** Centre of force is at the centre of ellipse)
16. A particle acted on by a central attraction force $\frac{\mu}{r^7}$ is projected with a velocity $\sqrt{\frac{\mu}{3a^6}}$ at a angle of 30° with its initial distance a from the centre of force. Show that the orbit is $r^2 = 2a^2 \cos 2\theta$.
17. A particle acted on by a central attraction force $\frac{\mu}{r^5}$ is projected with a velocity $\sqrt{\frac{\mu}{2a^4}}$ at a angle of α with its initial distance a from the centre of force. Show that the orbit is a circle through centre of force O and of diameter $a\operatorname{cosec}\alpha$.

§ 5. Apses – An apse *is a point on the central orbit at which the radius vector is either maximum or minimum. The distance between an apse and the centre of force is known as* **Apsidal distance** *and the angle subtended at the centre of force by two apsidal distances is known as* **Apsidal angle.** For example the motion of the earth moving round the sun is an elliptical motion, the ends of major axis are the two apses, one nearer to the sun (sun at the one of the focus) is known as **perihelion** and the farthest point is known as **aphelion.**

The apsidal angle in case of earth motion round the sun is 180^0. In circle all the points on the circumference are apses when pole is at origin.

Since at apse the radius vector (r) is maximum or minimum, therefore, we have

$$\frac{dr}{d\theta}=0 \Rightarrow \frac{du}{d\theta}=0 \quad \text{... (i)} \left(\text{As}\, u=\frac{1}{r} \text{ and } \frac{du}{d\theta}=-\frac{1}{r^2}\frac{dr}{d\theta}\right)$$

We know $\frac{1}{p^2}=u^2+\left(\frac{du}{d\theta}\right)^2$ ...(ii)

In view of (i) and (ii) we get $\frac{1}{p^2}=u^2=\frac{1}{r^2}$

$\Rightarrow \quad p=r$ ... (iii)

We know that $p=r\sin\phi$, using (iii) we get $\phi=90°$

Hence at an apse $p=r$ or $\phi=90^0$, p is the length of perpendicular from centre of force O upon the tangent and ϕ is the angle between radius vector and the tangent. $\phi=90^0 \Rightarrow$ radius vector and the tangent (direction of velocity) are perpendicular to each other. **Thus at an apse the velocity is perpendicular to radius vector.**

§ 6. Velocity in Some Standard Forms under Central Forces. (i) **Velocity at an Apse.** If the particle is projected with a velocity v_0 from an apse at a distance $\boldsymbol{a}$ from the centre of the force, then by virtue of apse $p = a$

For central force $vp=h \Rightarrow v_0 a=h \Rightarrow v_0=\frac{h}{a}$ is the velocity at an apse.

(ii) **Velocity on a circle.** When a particle describes a circle of radius a about the central force f directed towards the centre of the circle, the velocity at any point on its path is known as **Velocity on a circle.**

Let v_0 be the velocity of the particle which describes a circle of radius a. Then we have

$$r=a \Rightarrow u=\frac{1}{a} \quad \text{... (i) } (a \text{ is a radius of circle which is constant})$$

$$\Rightarrow \quad \frac{du}{d\theta}=0 \text{ and } \frac{d^2u}{d\theta^2}=0 \quad \text{... (ii)}$$

For central force $vp=h \Rightarrow v_0 a=h$ ... (iii)

($p = a$, for pole at the centre of the circle)

We know that for central orbit

$$f = h^2u^2\left(\frac{d^2u}{d\theta^2}+u\right)=(v_0a)^2\frac{1}{a^2}\left(0+\frac{1}{a}\right)=\frac{v_0^2}{a} \qquad \text{(using (ii) and (iii))}$$

$$\Rightarrow \quad v_0^2 = af \Rightarrow v_0 = \sqrt{af}$$

Hence $\sqrt{af}$ is the velocity of a particle on a circle of radius a with central force f directed towards the centre of the circle.

(iii) **Velocity from infinity.** *When a particle is at rest at infinity and moves under the influence of central force, it acquires a velocity to any point, is known as* **Velocity from infinity at that point.**

Let v_∞ be the velocity acquired by the particle at a distance ***a*** *from the centre of force,* moving under *central force* ***f***, which is at rest at infinity. We know that

$$\frac{f}{u^2} = h^2\left(\frac{d^2u}{d\theta^2}+u\right) \qquad \text{... (i)}$$

Multiplying both sides of (i) by $2\frac{du}{d\theta}$ and then integrate the result w.r. t. θ

$$\int 2\frac{du}{d\theta}\frac{f}{u^2}d\theta = \int h^2\left(2\frac{du}{d\theta}\frac{d^2u}{d\theta^2}+2\frac{du}{d\theta}u\right)d\theta$$

$$\Rightarrow \quad h^2\left(u^2+\left(\frac{du}{d\theta}\right)^2\right)=\int 2\frac{f}{u^2}du + C \qquad \text{... (ii)}$$

where C is an arbitrary constant.

For central force $vp = h \Rightarrow v^2 = \frac{h^2}{p^2} = h^2\left(u^2+\left(\frac{du}{d\theta}\right)^2\right)$ ... (iii)

In view of (ii) and (iii) we have

$$v^2 = \int 2\frac{f}{u^2}du + C \qquad \left(\text{using } u=\frac{1}{r}\Rightarrow \frac{du}{dr}=-\frac{1}{r^2}=-u^2 \Rightarrow \frac{du}{u^2}=-dr\right)$$

$$\Rightarrow \quad v^2 = -2\int f dr + C \qquad \text{...(iv)}$$

Where C is the constant of integration can be determined with the help of given condition of velocity of projection.

Since particle starts from rest at infinity and acquires the velocity v_0 at $r = a$.

$$\left[v^2\right]_0^{v_\infty} = -2\int_\infty^a f dr \Rightarrow v_\infty{}^2 = -2\int_\infty^a f dr = 2\int_a^\infty f dr \qquad \text{... (v)}$$

This is the required equation of velocity from infinity.

(iv) **Velocity of fall to the point of projection.** Let v_0 be the velocity acquired by the particle falls under *central repulsion force (–f)* at a distance *r* from the centre of force.

$$v_0^2 = 2\int_0^r f dr \qquad \text{... (vi)}$$

Important results to be used

1. At an apse $\frac{dr}{d\theta} = 0$ or $\frac{du}{d\theta} = 0$ or $p = r$.
2. Velocity at an Apse $v_0 = \frac{h}{a} \Rightarrow h = av_0$, Apse at a distance ***a*** from the centre of the force.
3. Velocity on a circle. $v_0 = \sqrt{af}$, circle of radius ***a*** about the central force *f* directed towards O.
4. Velocity from infinity $v_\infty^2 = 2\int_a^\infty f dr$, at a distance ***a*** from the centre of the force.
5. Velocity of fall to the point of projection $v_0^2 = 2\int_0^r f dr$, *central repulsion force f* at a distance *r* from O.

Type I. To find the Orbit, if the law of central force and velocity at apse v_0 are given.

For finding the orbit, we can use the algorithm as already elaborated in previous section illustrations. However, here another method is explained.

Algorithm to find the central orbit

Step 1. Find h^2 using $h^2 = p^2v^2 = a^2v_0^2$, where *a* is apsidal distance and v_0 is velocity at apse.

Step 2. Put the value of *f* in $f = \frac{h^2}{p^3}\frac{dp}{dr}$ and integrate the result.

Step 3. Substitute the value of h^2 in the result obtained in step 1 and apply the apse condition $p = a$, hence evaluate the integration constant.

Step 4. Use $\dfrac{1}{p^2} = \dfrac{1}{r^2} + \dfrac{1}{r^4}\left(\dfrac{dr}{d\theta}\right)^2$ and find $\dfrac{dr}{d\theta}$.

Step 5. Integrate the result of step 4 and apply apse condition $p = a$ at $\theta = 0$, hence evaluate the integration constant.

Note. *For the sake of convenience of readers some of the Examples are being solved by using both methods.*

ILLUSTRATIVE EXAMPLES

Example 1 A particle moves with a central acceleration $f = \dfrac{\mu}{r^5}$, if it is projected from an apse at a distance a from the origin with a velocity $= \sqrt{\dfrac{\mu}{2}\dfrac{1}{a^2}}$. Find the equation of the orbit.

Solution We have, according to question velocity at apse is

$$v_0 = \sqrt{\frac{\mu}{2}\frac{1}{a^2}} \Rightarrow v_0{}^2 = \frac{\mu}{2a^4} \qquad \text{... (1)}$$

Velocity at an Apse $v_0 = \dfrac{h}{a} \Rightarrow h^2 = a^2 v_0^2$

$$\Rightarrow \quad h^2 = a^2 \frac{\mu}{2a^4} \qquad \text{(using (1))}$$

$$\Rightarrow \quad h^2 = \frac{\mu}{2a^2} \qquad \text{... (2)}$$

We have $f = \dfrac{h^2}{p^3}\dfrac{dp}{dr} = \dfrac{\mu}{r^5} \quad \Rightarrow \quad \dfrac{h^2}{p^3}dp = \dfrac{\mu}{r^5}dr$

On integration we get

$$\int \frac{h^2}{p^3}dp = \int \frac{\mu}{r^5}dr$$

$$\Rightarrow \quad -\frac{h^2}{2p^2} = -\frac{\mu}{4r^4} + c$$

$$\Rightarrow \quad \frac{h^2}{p^2} = \frac{\mu}{2r^4} + c'$$ Using (2) we get,

$$\frac{\mu}{2a^2p^2} = \frac{\mu}{2r^4} + c' \qquad \text{... (3)}$$

We know that at apse $\frac{dr}{d\theta} = 0$ and $p = r = a$. Using these conditions in (4) we get,

$$\Rightarrow \quad \frac{\mu}{2a^2a^2} = \frac{\mu}{2a^4} + c' \quad \Rightarrow \quad c' = 0 \qquad \text{... (4)}$$

In view of (3) and (4)

$$\Rightarrow \quad \frac{\mu}{2a^2}\left(\frac{1}{r^2} + \frac{1}{r^4}\left(\frac{dr}{d\theta}\right)^2\right) = \frac{\mu}{2r^4}$$

$$\Rightarrow \quad \frac{1}{2a^2r^4}\left(\frac{dr}{d\theta}\right)^2 = \frac{1}{2r^4} - \frac{1}{2a^2r^2} \Rightarrow \left(\frac{dr}{d\theta}\right)^2 = a^2 - r^2$$

$$\Rightarrow \quad \frac{dr}{d\theta} = \pm\sqrt{(a^2 - r^2)}$$

Taking positive sign and on integration we get,

$$\int \frac{dr}{\sqrt{(a^2 - r^2)}} = \int d\theta + A \quad \Rightarrow \quad \sin^{-1}\frac{r}{a} = \theta + A$$

Initially $\theta = 0,\ r = a$ we get $A = \frac{\pi}{2}\left(\text{As } \sin^{-1} 1 = \frac{\pi}{2}\right)$

$$\Rightarrow \quad \sin^{-1}\frac{r}{a} = \theta + \frac{\pi}{2} \Rightarrow \frac{r}{a} = \sin\left(\theta + \frac{\pi}{2}\right)$$

$$\Rightarrow \quad r = a\cos\theta$$

2nd Method

We have, according to question velocity at apse is $v_0 = \sqrt{\frac{\mu}{2}}\frac{1}{a^2}$

$$\Rightarrow \quad v_0^2 = \frac{\mu}{2a^4} \qquad \text{... (1)}$$

Velocity at an Apse $v_0 = \frac{h}{a} \Rightarrow h^2 = a^2v_0^2$

$$\Rightarrow \quad h^2 = a^2 \frac{\mu}{2a^4} \qquad \text{(using (1))}$$

$$\Rightarrow \quad h^2 = \frac{\mu}{2a^2} \qquad \dots (2)$$

We have $f = \frac{\mu}{r^5} = \mu u^5$

$$\Rightarrow \quad \left(\frac{d^2u}{d\theta^2} + u\right) = \frac{f}{h^2u^2} = \frac{\mu u^5}{h^2u^2}$$

$$\Rightarrow \quad \left(\frac{d^2u}{d\theta^2} + u\right) = \frac{\mu}{h^2}u^3 \qquad \dots (3)$$

Substituting the value of h^2 from (2) in (3) we get

$$\left(\frac{d^2u}{d\theta^2} + u\right) = \frac{2a^2\mu}{\mu}u^3 = 2a^2u^3 \qquad \dots (4)$$

Multiplying both sides of (4) by $2\frac{du}{d\theta}$ and then integrate the result w.r. t. θ

$$2\frac{du}{d\theta}\frac{d^2u}{d\theta^2} + 2\frac{du}{d\theta}u = 2a^2u^3 2\frac{du}{d\theta}$$

$$\Rightarrow \quad \left(\frac{du}{d\theta}\right)^2 + u^2 = a^2u^4 + C$$

(We know that at apse $\frac{du}{d\theta} = 0$ and $p = r = a$ we get $C = 0$)

$$\Rightarrow \quad \left(\frac{du}{d\theta}\right)^2 = a^2u^4 - u^2 = u^2(a^2u^2 - 1) \;\Rightarrow\; \frac{du}{d\theta} = \pm u\sqrt{(a^2u^2 - 1)} \qquad \dots (5)$$

Taking positive sign and on integration we get,

$$\int \frac{du}{u\sqrt{(a^2u^2 - 1)}} = \int d\theta + A$$

$\sec^{-1} au = \theta + A$ Initially $\theta = 0$, $u = 1/a$ we get $A = 0$ (As $\sec^{-1} 1 = 0$).

$$\Rightarrow \quad au = \sec\theta \;\Rightarrow\; r = a\cos\theta$$

Example 2 A particle moves with a central acceleration $f = \frac{\mu}{r^3}$, if it is projected from an apse at a distance a from the origin with a velocity $= \frac{\sqrt{\mu}}{a}$. Find the equation of the orbit.

Solution We have, according to question velocity at apse is $v_0 = \frac{\sqrt{\mu}}{a}$

$$\Rightarrow \quad v_0^2 = \frac{\mu}{a^2} \qquad \text{... (1)}$$

Velocity at an Apse $v_0 = \frac{h}{a} \Rightarrow h^2 = a^2 v_0^2$

$$\Rightarrow \quad h^2 = a^2 \frac{\mu}{a^2} \qquad \text{(using (1))}$$

$$\Rightarrow \quad h^2 = \mu \qquad \text{... (2)}$$

We have $f = \frac{h^2}{p^3}\frac{dp}{dr} = \frac{\mu}{r^3} \Rightarrow \frac{h^2}{p^3}dp = \frac{\mu}{r^3}dr$

On integration we get $\int \frac{h^2}{p^3}dp = \int \frac{\mu}{r^3}dr$

$$\Rightarrow \quad -\frac{h^2}{2p^2} = -\frac{\mu}{2r^2} + c$$

$$\Rightarrow \quad \frac{h^2}{p^2} = \frac{\mu}{r^2} + c' \qquad \text{Using (2) we get,}$$

$$\frac{\mu}{p^2} = \frac{\mu}{r^2} + c' \qquad \text{... (3)}$$

We know that at apse $\frac{dr}{d\theta} = 0$ and $p = r = a$. Using these conditions in (4) we get,

$$\Rightarrow \quad \frac{\mu}{a^2} = \frac{\mu}{a^2} + c' \quad \Rightarrow \quad c' = 0 \qquad \text{... (4)}$$

In view of (3) and (4)

$$\Rightarrow \quad \mu\left(\frac{1}{r^2}+\frac{1}{r^4}\left(\frac{dr}{d\theta}\right)^2\right)=\frac{\mu}{r^2}$$

$$\Rightarrow \quad \left(\frac{dr}{d\theta}\right)^2=0$$

$$\Rightarrow \quad \frac{dr}{d\theta}=0$$

On integration we get, $r = \text{constant} = k$
Hence orbit is a circle with centre at origin.

2nd Method

We have, according to question velocity at apse is $v_0=\dfrac{\sqrt{\mu}}{a}$

$$\Rightarrow \quad v_0^2=\frac{\mu}{a^2} \qquad \ldots (1)$$

Velocity at an Apse $v_0=\dfrac{h}{a}\Rightarrow h^2=a^2v_0^2$

$$\Rightarrow \quad h^2=a^2\frac{\mu}{a^2} \qquad \text{(using (1))}$$

$$\Rightarrow \quad h^2=\mu \qquad \ldots (2)$$

We have $f=\dfrac{\mu}{r^3}=\mu u^3$

$$\Rightarrow \quad \left(\frac{d^2u}{d\theta^2}+u\right)=\frac{f}{h^2u^2}=\frac{\mu u^3}{h^2u^2}$$

$$\Rightarrow \quad \left(\frac{d^2u}{d\theta^2}+u\right)=\frac{\mu}{h^2}u \qquad \ldots (3)$$

Substituting the value of h^2 from (2) in (3) we get

$$\left(\frac{d^2u}{d\theta^2}+u\right)=\frac{\mu}{\mu}u=u \qquad \ldots (4)$$

Multiplying both sides of (4) by $2\dfrac{du}{d\theta}$ and then integrate the result w.r. t. θ

$$2\frac{du}{d\theta}\frac{d^2u}{d\theta^2}+2\frac{du}{d\theta}u=2u\frac{du}{d\theta}$$

$$\Rightarrow \quad \left(\frac{du}{d\theta}\right)^2+u^2=u^2+C$$

(We know that at apse $\dfrac{du}{d\theta}=0$ and $p=r=a$ we get $C=0$)

$$\Rightarrow \quad \left(\frac{du}{d\theta}\right)^2=0 \Rightarrow u=\text{constant}$$

$$\Rightarrow \quad r=\frac{1}{u}=\text{constant}=k$$

Hence orbit is a circle with centre at origin.

Example 3 A particle moves with a central acceleration $\mu\left(\dfrac{3a}{r^4}-\dfrac{2(a^2-b^2)}{r^5}\right)$ with the condition $a>b$ being projected from an apse at a distance $a+b$ from the pole with velocity $\dfrac{\sqrt{\mu}}{a+b}$. Show that the path of the particle is the curve $r=a+b\cos\theta$.

Solution We have, according to question velocity at apse is $v_0=\dfrac{\sqrt{\mu}}{a+b}$

$$\Rightarrow \quad v_0^2=\frac{\mu}{(a+b)^2} \qquad \dots (1)$$

Velocity at an Apse $v_0=\dfrac{h}{a+b}\Rightarrow h^2=(a+b)^2v_0^2$

$$\Rightarrow \quad h^2=(a+b)^2\frac{\mu}{(a+b)^2} \qquad \text{(using (1))}$$

$$\Rightarrow \quad h^2=\mu \qquad \dots (2)$$

We have $f = \frac{h^2}{p^3}\frac{dp}{dr} = \mu\left(\frac{3a}{r^4} - \frac{2(a^2 - b^2)}{r^5}\right)$

$$\Rightarrow \quad \frac{h^2}{p^3}dp = \mu\left(\frac{3a}{r^4} - \frac{2(a^2 - b^2)}{r^5}\right)dr$$

On integration we get $\int \frac{h^2}{p^3}dp = \int \mu\left(\frac{3a}{r^4} - \frac{2(a^2 - b^2)}{r^5}\right)dr$

$$\Rightarrow \quad -\frac{h^2}{2p^2} = \mu\left(-\frac{a}{r^3} + \frac{(a^2 - b^2)}{2r^4}\right) + c$$

$$\Rightarrow \quad \frac{h^2}{p^2} = \mu\left(\frac{2a}{r^3} - \frac{(a^2 - b^2)}{r^4}\right) + c'$$ Using (2) we get,

$$\Rightarrow \quad \frac{\mu}{p^2} = \mu\left(\frac{2a}{r^3} - \frac{(a^2 - b^2)}{r^4}\right) + c' \qquad \text{... (3)}$$

We know that at apse $\frac{dr}{d\theta} = 0$ and $p = r = a + b$. Using these conditions in (3) we get,

$$\Rightarrow \quad \frac{\mu}{(a+b)^2} = \mu\left(\frac{2a(a+b) - (a^2 - b^2)}{(a+b)^4}\right) + c' = \frac{\mu(a+b)^2}{(a+b)^4} + c'$$

$$\Rightarrow \quad c' = 0 \qquad \text{... (4)}$$

In view of (3) and (4)

$$\mu\left(\frac{1}{r^2} + \frac{1}{r^4}\left(\frac{dr}{d\theta}\right)^2\right) = \mu\left(\frac{2a}{r^3} - \frac{(a^2 - b^2)}{r^4}\right)$$

$$\Rightarrow \quad \left(\frac{dr}{d\theta}\right)^2 = 2ar - a^2 + b^2 - r^2 = b^2 - (r-a)^2$$

$$\Rightarrow \quad \frac{dr}{d\theta} = \sqrt{b^2 - (r-a)^2} \quad \int \frac{1}{\sqrt{b^2 - (r-a)^2}}dr = \int d\theta + A$$

On integration we get, $\sin^{-1}\dfrac{(r-a)}{b}=\theta+A$ ... (5)

Initially $\theta=0,\ r=a+b$ we get $A=\dfrac{\pi}{2}\left(As\ \sin^{-1}1=\dfrac{\pi}{2}\right)$.

$$\Rightarrow \quad \sin^{-1}\frac{(r-a)}{b}=\theta+\frac{\pi}{2}$$

$$\Rightarrow \quad \frac{(r-a)}{b}=\sin\left(\theta+\frac{\pi}{2}\right)=\cos\theta$$

$$\Rightarrow \quad r-a=b\cos\theta$$

$$\Rightarrow \quad r=a+b\cos\theta.$$

Example 4 A particle moves with a central acceleration $\mu(r^5-a^4r)$, being projected from an apse at a distance *a* from the pole with velocity $\sqrt{\dfrac{2\mu}{3}}a^3$. Show that the path of the particle is the curve $x^4+y^4=a^4$.

Solution We have, according to question velocity at apse is $v_0=\sqrt{\dfrac{2\mu}{3}}a^3$

$$\Rightarrow \quad {v_0}^2=\frac{2\mu}{3}a^6 \qquad \text{... (1)}$$

Velocity at an Apse $v_0=\dfrac{h}{a}\Rightarrow h^2=a^2{v_0}^2$

$$\Rightarrow \quad h^2=a^2\frac{2\mu}{3}a^6 \qquad \text{(using (1))}$$

$$\Rightarrow \quad h^2=\frac{2\mu}{3}a^8 \qquad \text{... (2)}$$

We have $f=\dfrac{h^2}{p^3}\dfrac{dp}{dr}=\mu(r^5-a^4r)\Rightarrow\dfrac{h^2}{p^3}dp=\mu(r^5-a^4r)dr$

On integration we get $\displaystyle\int\frac{h^2}{p^3}dp=\int\mu(r^5-a^4r)dr$

$$\Rightarrow \quad -\frac{h^2}{2p^2} = \mu\left(\frac{r^6}{6} - a^4\frac{r^2}{2}\right) + c$$

$$\Rightarrow \quad \frac{h^2}{p^2} = \mu\left(a^4 r^2 - \frac{r^6}{3}\right) + c' \text{ Using (2) we get,}$$

$$\Rightarrow \quad \frac{2\mu}{3}a^8\frac{1}{p^2} = \mu\left(a^4 r^2 - \frac{r^6}{3}\right) + c' \qquad \dots (3)$$

We know that at apse $\frac{dr}{d\theta} = 0$ and $p = r = a$. Using these conditions in (3) we get,

$$\Rightarrow \quad \frac{2\mu a^6}{3} = \frac{2\mu a^6}{3} + c' \quad \Rightarrow \quad c' = 0 \qquad \dots (4)$$

In view of (3) and (4)

$$\frac{2\mu}{3}a^8\left(\frac{1}{r^2} + \frac{1}{r^4}\left(\frac{dr}{d\theta}\right)^2\right) = \mu\left(a^4 r^2 - \frac{r^6}{3}\right)$$

$$\Rightarrow \quad \left(\frac{dr}{d\theta}\right)^2 = r^4\left(\frac{3r^2}{2a^4} - \frac{r^6}{2a^8} - \frac{1}{r^2}\right) = \frac{r^2}{2a^8}\left(3a^4r^4 - r^8 - 2a^8\right)$$

$$\Rightarrow \quad \frac{dr}{d\theta} = \frac{r}{a^4}\frac{1}{\sqrt{2}}\sqrt{\left(3a^4r^4 - r^8 - 2a^8\right)}$$

$$\int \frac{a^4}{r\sqrt{(2a^4 - r^4)(r^4 - a^4)}}\,dr = \frac{1}{\sqrt{2}}\int d\theta + A$$

On integration we get,

$$\frac{1}{2\sqrt{2}}\sin^{-1}\sqrt{2}\sqrt{\left(\frac{r^4 - a^4}{r^4}\right)} = \frac{1}{\sqrt{2}}\theta + A \qquad \dots (5)$$

Initially $\theta = 0$, $r = a$ we get $A = 0$.

$$\Rightarrow \quad \sqrt{2\left(\frac{r^4 - a^4}{r^4}\right)} = \sin 2\theta$$

$$\Rightarrow \quad 2\left(\frac{r^4 - a^4}{r^4}\right) = \sin^2 2\theta = 4\sin^2\theta\cos^2\theta$$

$$\Rightarrow \quad r^4 - a^4 = 4r^4 \sin^2\theta \cos^2\theta = 2r^2 \sin^2\theta r^2 \cos^2\theta$$

Using $x = r\cos\theta$, $y = r\sin\theta$ and $r^2 = x^2 + y^2$, we get,

$$\Rightarrow \quad (x^2 + y^2)^2 - a^4 = 2x^2y^2$$

$$\Rightarrow \quad x^4 + y^4 = a^4.$$

Example 5 A particle moves with a central repulsion which varies inversely as the cube of the distance, if it is projected from an apse at a distance *a* from the origin with a velocity *V*. Show that the equation to the path is $r \cos p\theta = a$ and the angle described in time *t* is $\frac{1}{p}\tan^{-1}\left(\frac{pV}{a}t\right)$, where

$$p^2 = \frac{\mu + a^2V^2}{a^2V^2}.$$

Solution We know that

$$\text{Velocity at an Apse } V = \frac{h}{a} \Rightarrow h^2 = a^2V^2$$

$$\Rightarrow \qquad h^2 = a^2V^2 \qquad \dots (1)$$

We have $f = -\frac{\mu}{r^3} = -\mu u^3$ f (is repulsive in nature therefore, it is negative)

$$\Rightarrow \quad \left(\frac{d^2u}{d\theta^2} + u\right) = -\frac{f}{h^2u^2} = -\frac{\mu u^3}{a^2V^2u^2}$$

$$\Rightarrow \quad \left(\frac{d^2u}{d\theta^2} + u\right) = -\frac{\mu}{a^2V^2}u \qquad \dots (2)$$

$$\Rightarrow \quad \frac{d^2u}{d\theta^2} + \left(1 + \frac{\mu}{a^2V^2}\right)u = 0 \qquad \dots (3)$$

Using the operator $\frac{d}{d\theta} = D \Rightarrow \frac{d^2}{d\theta^2} = D^2$ we get,

Auxiliary equation of (3) will be

$$D^2 + \left(1 + \frac{\mu}{a^2V^2}\right) = 0 \Rightarrow D = \pm i\sqrt{\frac{\mu + a^2V^2}{a^2V^2}} \qquad \dots (4)$$

In view of (4) the solution of (3) is

$$u = A\cos p\theta + B\sin p\theta \qquad \text{... (5)}$$

$$\Rightarrow \quad \frac{du}{d\theta} = p(-A\sin p\theta + B\cos p\theta) \qquad \text{... (6)}$$

Initially at apse $\theta = 0, u = \frac{1}{a}$ and $\frac{du}{d\theta} = 0$. Using these conditions in (5) and (6) we get, $\frac{1}{a} = A$ and $0 = B \Rightarrow A = \frac{1}{a} B = 0$. Putting these values in (5) we get, $u = \frac{1}{a}\cos p\theta$

$$\Rightarrow \quad \frac{1}{r} = \frac{1}{a}\cos p\theta$$

$$\Rightarrow \quad r\cos p\theta = a$$

2nd part. We know that

$$r^2\dot{\theta} = h = aV$$

$$\Rightarrow \quad \dot{\theta} = \frac{aV}{r^2} = aV\left(\frac{\cos p\theta}{a}\right)^2 = \frac{V}{a}\cos^2 p\theta$$

$$\Rightarrow \quad \frac{d\theta}{dt} = \frac{V}{a}\cos^2 p\theta \Rightarrow \int_0^t \frac{V}{a}dt = \int_0^\theta \sec^2 p\theta d\theta$$

$$\Rightarrow \quad \frac{V}{a}t = \frac{\tan p\theta}{p}$$

$$\Rightarrow \quad t = \frac{1}{p}\tan^{-1}\left(\frac{pV}{a}t\right).$$

Type II. To find the Orbit, if the velocity at apse is in term of the velocity from infinity—

Note. *Find v_∞ using $v_\infty^2 = 2\int_a^\infty f dr$, then follow the algorithm already explained in type I.*

ILLUSTRATIVE EXAMPLES

Example 1 A particle is projected from an apse at a distance *a* with a velocity from infinite under the effect of same central force $f = \frac{\mu}{r^{2n+3}}$. Prove that the equation of orbit is $r^n = a^n \cos n\theta$.

Solution We know that

Velocity from infinity $v_\infty^2 = 2\int_a^\infty f dr$,

$$\Rightarrow \quad v_\infty^2 = 2\int_a^\infty \frac{\mu}{r^{2n+3}} dr, \qquad \ldots (1)$$

$$\Rightarrow \quad v_\infty^2 = \frac{1}{(n+1)} \frac{\mu}{a^{2n+2}}$$

According to question velocity at apse $V_1 = v_\infty$

$$\Rightarrow \quad V_1^2 = \frac{1}{(n+1)} \frac{\mu}{a^{2n+2}} \qquad \ldots (2)$$

Velocity at an Apse $V_1 = \frac{h}{a} \Rightarrow h^2 = a^2 V_1^2$

$$\Rightarrow \quad h^2 = a^2 V_1^2 = a^2 \frac{1}{(n+1)} \frac{\mu}{a^{2n+2}} \qquad \text{(using (2))}$$

$$\Rightarrow \quad h^2 = \frac{1}{(n+1)} \frac{\mu}{a^{2n}} \qquad \ldots (3)$$

We have $f = \frac{h^2}{p^3} \frac{dp}{dr} = \frac{\mu}{r^{2n+3}}$ Using (3)

$$\Rightarrow \quad \frac{1}{p^3} \frac{1}{(n+1)} \frac{\mu}{a^{2n}} dp = \frac{\mu}{r^{2n+3}} dr$$

On integration we get $\frac{1}{(n+1)a^{2n}} \int \frac{1}{p^3} dp = \int \frac{1}{r^{2n+3}} dr$

$$\Rightarrow \quad -\frac{1}{(n+1)a^{2n}} \frac{1}{2p^2} = -\frac{1}{(2n+2)r^{2n+2}} + c$$

$$\Rightarrow \quad \frac{1}{a^{2n}p^2} = \frac{1}{r^{2n+2}} + c' \qquad \dots (4)$$

We know that at apse $\frac{dr}{d\theta} = 0$ and $p = r = a$. Using these conditions in (4) we get,

$$\Rightarrow \quad \frac{2\mu}{a^{2n+2}} = \frac{\mu}{a^{2n+2}} + c \Rightarrow c' = 0$$

$$\Rightarrow \quad \frac{1}{a^{2n}}\left(\frac{1}{r^2} + \frac{1}{r^4}\left(\frac{dr}{d\theta}\right)^2\right) = \frac{1}{r^{2n+2}}$$

$$\Rightarrow \quad \frac{1}{a^{2n}}\frac{1}{r^4}\left(\frac{dr}{d\theta}\right)^2 = \frac{1}{r^{2n+2}} - \frac{1}{a^{2n}}\frac{1}{r^2}$$

$$\Rightarrow \quad \left(\frac{dr}{d\theta}\right)^2 = \frac{\left(a^{2n} - r^{2n}\right)}{r^{2n-2}}$$

$$\Rightarrow \quad \frac{dr}{d\theta} = \pm\sqrt{\frac{\left(a^{2n} - r^{2n}\right)}{r^{2n-2}}}$$

Taking positive sign and on integration we get,

$$\int \frac{r^{n-1}dr}{\sqrt{\left(a^{2n} - r^{2n}\right)}} = \int d\theta + A' \qquad \dots (5)$$

Put $r^n = a^n \sin\alpha \Rightarrow nr^{n-1}dr = a^n \cos\alpha\, d\alpha$

$$\Rightarrow \quad \alpha = \sin^{-1}\left(\frac{r^n}{a^n}\right) \qquad \dots (6)$$

$$\Rightarrow \quad \int \frac{1}{n} d\alpha = \int d\theta + A'$$

$$\Rightarrow \quad \alpha = n\theta + A \qquad \text{(using (6))}$$

$$\Rightarrow \quad \sin^{-1}\left(\frac{r^n}{a^n}\right) = n\theta + A.$$

Initially $\theta = 0, r = a,$ we get $A = \frac{\pi}{2}$ $\left(\text{As } \sin^{-1} 1 = \frac{\pi}{2}\right).$

$$\Rightarrow \quad \frac{r^n}{a^n} = \sin\left(n\theta + \frac{\pi}{2}\right)$$

$$\Rightarrow \quad r^n = a^n \cos n\theta.$$

Example 2. A particle is projected at right angle at a distance *a* with a velocity from infinite under the effect of same central force $f = \dfrac{\mu}{r^n}$. Prove that the equation of orbit is $r^{(n-3)/2} = a^{(n-3)/2} \cos \dfrac{(n-3)}{2}\theta$.

Solution We know that

Velocity from infinity $v_\infty^2 = 2\int_a^\infty f dr$,

$$\Rightarrow \quad v_\infty^2 = 2\int_a^\infty \frac{\mu}{r^n} dr, \qquad \text{... (1)}$$

$$\Rightarrow \quad v_\infty^2 = \frac{2}{(n-1)} \frac{\mu}{a^{n-1}}$$

Velocity is perpendicular to radius vector i.e. the point is apse.

According to question Velocity at apse $V_1 = v_\infty$

$$\Rightarrow \quad V_1^2 = \frac{2}{(n-1)} \frac{\mu}{a^{n-1}} \qquad \text{... (2)}$$

Velocity at an Apse $V_1 = \dfrac{h}{a} \Rightarrow h^2 = a^2 V_1^2$

$$\Rightarrow \quad h^2 = a^2 V_1^2 = a^2 \frac{2}{(n-1)} \frac{\mu}{a^{n-1}} \qquad \text{(using (2))}$$

$$\Rightarrow \quad h^2 = \frac{2}{(n-1)} \frac{\mu}{a^{n-3}} \qquad \text{... (3)}$$

We have $f = \dfrac{h^2}{p^3} \dfrac{dp}{dr} = \dfrac{\mu}{r^n}$ Using (3)

$$\Rightarrow \quad \frac{1}{p^3} \frac{2}{(n-1)} \frac{\mu}{a^{n-3}} dp = \frac{\mu}{r^n} dr$$

On integration we get $\dfrac{2}{(n-1)a^{n-3}}\int \dfrac{1}{p^3}dp = \int \dfrac{1}{r^n}dr$

$$\Rightarrow \quad -\frac{1}{(n-1)a^{n-3}}\frac{1}{p^2} = -\frac{1}{(n-1)r^{n-1}} + c$$

$$\Rightarrow \quad \frac{1}{a^{n-3}}\frac{1}{p^2} = \frac{1}{r^{n-1}} + c' \qquad \text{... (4)}$$

We know that at apse $\dfrac{dr}{d\theta} = 0$ and $p = r = a$. Using these conditions in (4) we get,

$$\Rightarrow \quad \frac{\mu}{a^{n-1}} = \frac{\mu}{a^{n-1}} + c' \Rightarrow c' = 0$$

$$\Rightarrow \quad \frac{1}{a^{n-3}}\left(\frac{1}{r^2} + \frac{1}{r^4}\left(\frac{dr}{d\theta}\right)^2\right) = \frac{1}{r^{n-1}}$$

$$\Rightarrow \quad \frac{1}{a^{n-3}}\frac{1}{r^4}\left(\frac{dr}{d\theta}\right)^2 = \frac{1}{r^{n-1}} - \frac{1}{a^{n-3}}\frac{1}{r^2}$$

$$\Rightarrow \quad \left(\frac{dr}{d\theta}\right)^2 = \frac{\left(a^{n-3} - r^{n-3}\right)}{r^{n-5}}$$

$$\Rightarrow \quad \frac{dr}{d\theta} = \pm\sqrt{\frac{\left(a^{n-3} - r^{n-3}\right)}{r^{n-5}}}$$

Taking positive sign and on integration we get,

$$\int \frac{r^{\frac{(n-3)}{2}-1}\,dr}{\sqrt{\left(a^{n-3} - r^{n-3}\right)}} = \int d\theta + A' \qquad \text{... (5)}$$

Put $r^{\frac{(n-3)}{2}} = a^{\frac{(n-3)}{2}}\sin\alpha \Rightarrow \dfrac{(n-3)}{2}r^{\frac{(n-3)}{2}-1}dr = a^{\frac{(n-3)}{2}}\cos\alpha\, d\alpha$

$$\Rightarrow \quad \alpha = \sin^{-1}\left(\frac{r^{\frac{(n-3)}{2}}}{a^{\frac{(n-3)}{2}}}\right) \qquad \text{... (6)}$$

$$\Rightarrow \quad \int \frac{2}{n-3}\,d\alpha = \int d\theta + A'$$

$$\Rightarrow \quad \alpha = \frac{n-3}{2}\theta + A \qquad \text{(using (6))}$$

$$\Rightarrow \quad \sin^{-1}\left(\frac{r^{\frac{(n-3)}{2}}}{a^{\frac{(n-3)}{2}}}\right) = \frac{n-3}{2}\theta + A. \text{ Initially } \theta = 0,\ r = a, \text{ we get } A = \frac{\pi}{2}$$

$$\left(\text{As } \sin^{-1} 1 = \frac{\pi}{2}\right)$$

$$\Rightarrow \quad \left(\frac{r^{\frac{(n-3)}{2}}}{a^{\frac{(n-3)}{2}}}\right) = \sin\left(\frac{n-3}{2}\theta + \frac{\pi}{2}\right)$$

$$\Rightarrow \quad r^{\frac{(n-3)}{2}} = a^{\frac{(n-3)}{2}} \cos\frac{(n-3)}{2}\theta.$$

Type III To find the Orbit, if the velocity at apse is in term of the velocity on a circle—

Note. *Find* v_0 *using* $v_0 = \sqrt{af}$, *and then follow the algorithm already explained in previous type.*

ILLUSTRATIVE EXAMPLES

Example 1 A particle moves with a central acceleration which varies inversely as the cube of the distance, if it is projected from an apse at a distance a from the origin with a velocity which is $\sqrt{2}$ times the velocity for a circle of radius a with the same central force. Show that the equation to the path is $r\cos\frac{\theta}{\sqrt{2}} = a$.

Solution We know that $v_0 = \sqrt{af}$ =Velocity on a circle of radius a

$$\Rightarrow \quad v_0 = \sqrt{a\frac{\mu}{a^3}} = \frac{\sqrt{\mu}}{a} \qquad \text{... (1)}$$

According to question velocity at apse $=\sqrt{2}$ times the velocity for a circle of radius a.

$$V_1 = \sqrt{2}v_0 = \sqrt{2}\frac{\sqrt{\mu}}{a}$$

$$\Rightarrow \quad V_1^2 = \frac{2\mu}{a^2} \qquad \dots (2)$$

Velocity at an Apse $V_1 = \frac{h}{a} \Rightarrow h^2 = a^2V_1^2$

$$\Rightarrow \quad h^2 = a^2V_1^2 = a^2\frac{2\mu}{a^2} \qquad \text{(using (2))}$$

$$\Rightarrow \quad h^2 = 2\mu \qquad \dots (3)$$

We have $f = \frac{\mu}{r^3} = \mu u^3$

$$\Rightarrow \quad \left(\frac{d^2u}{d\theta^2} + u\right) = \frac{f}{h^2u^2} = \frac{\mu u^3}{h^2u^2}$$

$$\Rightarrow \quad \left(\frac{d^2u}{d\theta^2} + u\right) = \frac{\mu}{h^2}u \qquad \dots (4)$$

Substituting the value of h^2 from (3) in (4) we get $\left(\frac{d^2u}{d\theta^2} + u\right) = \frac{\mu}{2\mu}u = \frac{u}{2}$

$$\Rightarrow \quad \frac{d^2u}{d\theta^2} + \frac{u}{2} = 0 \qquad \dots (5)$$

Using the operator $\frac{d}{d\theta} = D \Rightarrow \frac{d^2}{d\theta^2} = D^2$ we get,

Auxiliary equation of (5) will be $D^2 + \frac{1}{2} = 0$

$$\Rightarrow \quad D = \pm\frac{i}{\sqrt{2}} \qquad \dots (6)$$

In view of (6) the solution of (5) is

$$u = A\cos\frac{\theta}{\sqrt{2}} + B\sin\frac{\theta}{\sqrt{2}} \quad \ldots (7)$$

$$\Rightarrow \quad \frac{du}{d\theta} = \frac{1}{\sqrt{2}}\left(-A\sin\frac{\theta}{\sqrt{2}} + B\cos\frac{\theta}{\sqrt{2}}\right) \quad \ldots (8)$$

Initially at apse $\theta = 0$, $u = \frac{1}{a}$ and $\frac{du}{d\theta} = 0$. Using these conditions in (7) and (8) we get, $\frac{1}{a} = A$ and $0 = B \Rightarrow A = \frac{1}{a}$ $B = 0$. Putting these values in (7) we get, $u = \frac{1}{a}\cos\frac{\theta}{\sqrt{2}}$

$$\Rightarrow \quad \frac{1}{r} = \frac{1}{a}\cos\frac{\theta}{\sqrt{2}}$$

$$\Rightarrow \quad r\cos\frac{\theta}{\sqrt{2}} = a$$

2nd Method

We know that $v_0 = \sqrt{af}$ =Velocity on a circle of radius a

$$\Rightarrow \quad v_0 = \sqrt{a\frac{\mu}{a^3}} = \frac{\sqrt{\mu}}{a} \quad \ldots (1)$$

According to question velocity at apse $= \sqrt{2}$ times the velocity for a circle of radius a.

$$V_1 = \sqrt{2}v_0 = \sqrt{2}\frac{\sqrt{\mu}}{a}$$

$$\Rightarrow \quad V_1^2 = \frac{2\mu}{a^2} \quad \ldots (2)$$

Velocity at an Apse $V_1 = \frac{h}{a} \Rightarrow h^2 = a^2V_1^2$

$$\Rightarrow \quad h^2 = a^2V_1^2 = a^2\frac{2\mu}{a^2} \qquad \text{(using (2))}$$

$$\Rightarrow \quad h^2 = 2\mu \quad \ldots (3)$$

We have $f=\frac{h^2}{p^3}\frac{dp}{dr}=\frac{\mu}{r^3}\Rightarrow\frac{h^2}{p^3}dp=\frac{\mu}{r^3}dr$

On integration we get $\int\frac{h^2}{p^3}dp=\int\frac{\mu}{r^3}dr$

$$\Rightarrow\quad -\frac{h^2}{2p^2}=-\frac{\mu}{2r^2}+c$$

$$\Rightarrow\quad \frac{h^2}{p^2}=\frac{\mu}{r^2}+c' \text{ Using (3) we get,}$$

$$\frac{2\mu}{p^2}=\frac{\mu}{r^2}+c' \qquad \ldots (4)$$

We know that at apse $\frac{dr}{d\theta}=0$ and $p=r=a$. Using these conditions in (4) we get,

$$\Rightarrow\quad \frac{2\mu}{a^2}=\frac{\mu}{a^2}+c'\Rightarrow c'=\frac{\mu}{a^2}$$

$$\Rightarrow\quad 2\mu\left(\frac{1}{r^2}+\frac{1}{r^4}\left(\frac{dr}{d\theta}\right)^2\right)=\mu\left(\frac{1}{r^2}+\frac{1}{a^2}\right)$$

$$\Rightarrow\quad \frac{1}{r^4}\left(\frac{dr}{d\theta}\right)^2=\frac{1}{2a^2}-\frac{1}{2r^2}\Rightarrow\left(\frac{dr}{d\theta}\right)^2=r^2\frac{(r^2-a^2)}{2a^2}$$

$$\Rightarrow\quad \frac{dr}{d\theta}=\pm\frac{r}{a\sqrt{2}}\sqrt{(r^2-a^2)}$$

Taking positive sign and on integration we get,

$$\Rightarrow\quad \int\frac{dr}{\frac{r}{a}\sqrt{\left(\left(\frac{r}{a}\right)^2-1\right)}}=\frac{\theta}{\sqrt{2}}+A$$

$$\Rightarrow\quad \sec^{-1}\frac{r}{a}=\frac{\theta}{\sqrt{2}}+A$$

Initially $\theta = 0, r = a$ we get A (As $\sec^{-1} 1 = 0$).

$$\Rightarrow \quad \sec^{-1}\frac{r}{a} = \frac{\theta}{\sqrt{2}} \Rightarrow \frac{r}{a} = \sec\frac{\theta}{\sqrt{2}}$$

$$\Rightarrow \quad r\cos\frac{\theta}{\sqrt{2}} = a$$

Example 2 A particle is moving in a curve under a central attraction force so that its velocity at any point is equal to that in a circle of the same distance and under the same attraction law of force. Show that the law of force is inverse cube and the path is equiangular spiral.

Solution We know that

$$v = \sqrt{rf} = \text{Velocity on a circle of radius } r$$

$$\Rightarrow \quad v^2 = rf \qquad \ldots (1)$$

According to question, velocity at any point at a distance r on the orbit = the velocity for a circle of radius r.

For central orbit we have $h^2 = p^2v^2$ and $f = \dfrac{h^2}{p^3}\dfrac{dp}{dr}$

$$\Rightarrow \quad h^2 = p^2rf \ \ldots(2) \text{ (using (1)) } f = \frac{h^2}{p^3}\frac{dp}{dr} = \frac{p^2rf}{p^3}\frac{dp}{dr} \text{ (using (2))}$$

$$\Rightarrow \quad \frac{r}{p}\frac{dp}{dr} = 1 \qquad \ldots (3)$$

On integration we get,

$$\Rightarrow \quad \int\frac{dp}{p} = \int\frac{dr}{r} \Rightarrow \log p = \log r + \log c$$

$$\Rightarrow \quad p = cr \Rightarrow p^2 = Cr^2 \Rightarrow \frac{1}{p^2} = \frac{1}{Cr^2} \qquad \ldots (4)$$

$$\Rightarrow \quad \frac{1}{r^2} + \frac{1}{r^4}\left(\frac{dr}{d\theta}\right)^2 = \frac{1}{Cr^2}$$

$$\Rightarrow \quad \left(\frac{dr}{d\theta}\right)^2 = \frac{C-1}{C}r^2 \Rightarrow \frac{dr}{d\theta} = \sqrt{\frac{C-1}{C}}\,r$$

$$\Rightarrow \quad \int \frac{dr}{r} = k\int d\theta + \log A \text{ where } k = \sqrt{\frac{C-1}{C}}$$

$$\Rightarrow \quad \log r = k\theta + \log A$$

$\Rightarrow \quad r = Ae^{k\theta}$. Hence path is equiangular spiral

Now differentiating (4) we get

$$\frac{-2}{p^3}\frac{dp}{dr} = \frac{-2}{r^3} \Rightarrow \frac{1}{p^3}\frac{dp}{dr} = \frac{1}{r^3} \qquad \ldots (5)$$

We know that $f = \dfrac{h^2}{p^3}\dfrac{dp}{dr} = \dfrac{h^2}{r^3}$ (using (5))

Hence law of force is inverse cube.

Type IV. To find the Orbit, if the velocity at apse is in term of the velocity acquired in falling to the point of projection from the centre by the central repulsive force.

Note. *Find* v_0 *using* $v_0^2 = 2\int_0^a f dr$, *then follow the algorithm already explained in type I.*

ILLUSTRATIVE EXAMPLES

Example 1 A particle is moving under a constant force from the centre is projected in the direction perpendicular to the radius vector at a distance *a* with the velocity acquired in falling to the point of projection from the centre. Show that its path is $\left(\frac{a}{r}\right)^3 = \cos^2\frac{3\theta}{2}$

Solution We know that

Velocity of projection v_0 = Velocity under *central repulsion force f* at a distance a from the centre of force.

$$\Rightarrow \quad v_0^2 = -2\int_0^a f dr$$

The law of force is constant repulsive in nature $f = -k$

$$\Rightarrow \quad v_0^2 = 2\int_0^a k dr, \quad \Rightarrow \quad v_0^2 = 2ka \qquad \ldots (1)$$

Velocity is perpendicular to radius vector i. e. the point is apse.

According to question Velocity at apse $V_1 = v_0$

$$\Rightarrow \quad V_1^2 = 2ka \qquad \ldots (2)$$

Velocity at an Apse $V_1 = \frac{h}{a} \Rightarrow h^2 = a^2 V_1^2$

$$\Rightarrow \quad h^2 = a^2 V_1^2 = 2ka^3 \qquad \text{(using (2))}$$

$$\Rightarrow \quad h^2 = 2ka^3 \qquad \ldots (3)$$

We have $f = \frac{h^2}{p^3}\frac{dp}{dr} = -k$ (using (3))

$$\Rightarrow \quad \frac{1}{p^3} 2ka^3 dp = -kdr$$

On integration we get $2a^3 \int \frac{1}{p^3} dp = -\int dr$

$$\Rightarrow \quad a^3 \frac{1}{p^2} = r + c' \qquad \ldots (4)$$

We know that at apse $\frac{dr}{d\theta} = 0$ and $p = r = a$. Using these condition in (4) we get,

$$\Rightarrow \quad a = a + c' \Rightarrow c' = 0$$

$$\Rightarrow \quad a^3 \left(\frac{1}{r^2} + \frac{1}{r^4} \left(\frac{dr}{d\theta} \right)^2 \right) = r$$

$$\Rightarrow \quad \frac{1}{r^4} \left(\frac{dr}{d\theta} \right)^2 = \frac{r}{a^3} - \frac{1}{r^2} \Rightarrow \left(\frac{dr}{d\theta} \right)^2 = \frac{(r^5 - r^2 a^3)}{a^3}$$

$$\Rightarrow \quad \frac{dr}{d\theta} = \pm \sqrt{\frac{(r^5 - r^2 a^3)}{a^3}}$$

Taking positive sign and on integration we get,

$$\int \frac{a^{3/2} dr}{r\sqrt{(r^3 - a^3)}} = \int d\theta + A' \qquad \ldots (5)$$

Put $r^{\frac{3}{2}} = a^{\frac{3}{2}} \sec\alpha \Rightarrow \frac{3}{2} r^{\frac{1}{2}} dr = a^{\frac{3}{2}} \sec\alpha \tan\alpha \, d\alpha$

$$\Rightarrow \quad \alpha = \sec^{-1}\left(\frac{r^{\frac{3}{2}}}{a^{\frac{3}{2}}}\right) \qquad \ldots (6)$$

$$\Rightarrow \quad \int \frac{a^{3/2}\, 2a^{3/2} \sec\alpha \tan\alpha \, d\alpha}{3a^{3/2} \sec\alpha \, a^{3/2} \tan\alpha} = \int d\theta + A'$$

$$\Rightarrow \quad \alpha = \frac{3}{2}\theta + A \qquad \text{(using (6))}$$

$$\Rightarrow \quad \sec^{-1}\left(\frac{r^{\frac{3}{2}}}{a^{\frac{3}{2}}}\right) = \frac{3}{2}\theta + A.$$

Initially $\theta = 0, r = a$, we get $A = 0$ $\left(\text{As } \sec^{-1} 1 = 0\right)$

$$\Rightarrow \quad \left(\frac{r^{\frac{3}{2}}}{a^{\frac{3}{2}}}\right) = \sec\frac{3}{2}\theta \Rightarrow \left(\frac{r}{a}\right)^3 = \sec^2\frac{3}{2}\theta.$$

$$\Rightarrow \quad \left(\frac{a}{r}\right)^3 = \cos^2\frac{3\theta}{2}.$$

Example 2 A particle is moving by a central repulsive force which varies as the nth power of the distance, if the velocity at any point of the path is equal to that which would be acquired in falling from the centre to the point. Show that its path is $\sin^{-1}\left(\frac{r^{\frac{(n-1)}{2}}}{\sqrt{A}}\right) = \frac{(n-1)}{2}\theta + c'''$.

Solution We know that velocity at any point of the path v = Velocity under *central repulsion force f* at a distance a from the centre of force.

$$\Rightarrow \quad v^2 = -2\int_0^r f dr$$

The law of force is constant repulsive in nature $f = -\mu r^n$

$$\Rightarrow \quad v^2 = 2\int_0^r \mu r^n dr, \quad \Rightarrow \quad v^2 = \frac{2\mu r^{n+1}}{(n+1)} \qquad \ldots (1)$$

According to question Velocity at any point on the path $V_1 = v$

$$\Rightarrow \quad V_1^2 = \frac{2\mu r^{n+1}}{(n+1)} \qquad \ldots (2)$$

Velocity at any point $V_1 = \frac{h}{p} \Rightarrow h^2 = p^2 V_1^2$ (using (2))

$$\Rightarrow \quad h^2 = p^2 \frac{2\mu r^{n+1}}{(n+1)} \qquad \ldots (3)$$

We have $f = \frac{h^2}{p^3}\frac{dp}{dr} = -\mu r^n$ Using (3)

$$\Rightarrow \quad \frac{1}{p^3} p^2 \frac{2\mu r^{n+1}}{(n+1)}\frac{dp}{dr} = -\mu r^n \quad \frac{2}{(n+1)}\frac{dp}{p} = -\frac{dr}{r}$$

On integration we get $-2\int \frac{1}{p} dp = (n+1)\int \frac{dr}{r}$

$$\Rightarrow \quad -2\log p = (n+1)\log r + c' \qquad \ldots (4)$$

$$\Rightarrow \quad \frac{1}{p^2} = \frac{A}{r^{n+1}}$$

$$\Rightarrow \quad \left(\frac{1}{r^2} + \frac{1}{r^4}\left(\frac{dr}{d\theta}\right)^2\right) = \frac{A}{r^{n+1}}$$

$$\Rightarrow \quad \frac{1}{r^4}\left(\frac{dr}{d\theta}\right)^2 = \frac{A}{r^{n+1}} - \frac{1}{r^2} \Rightarrow \left(\frac{dr}{d\theta}\right)^2 = \frac{\left(A - r^{n-1}\right)}{r^{n-3}}$$

$$\Rightarrow \quad \frac{dr}{d\theta} = \pm\sqrt{\frac{\left(A - r^{n-1}\right)}{r^{n-3}}}$$

Taking positive sign and on integration we get,

$$\int \frac{r^{(n-3)/2}\,dr}{\sqrt{\left(A - r^{n-1}\right)}} = \int d\theta + A' \qquad \ldots (5)$$

Put $r^{\frac{(n-1)}{2}} = \sqrt{A}\sin\alpha \Rightarrow \frac{(n-1)}{2} r^{\frac{(n-3)}{2}}\, dr = \sqrt{A}\cos\alpha\, d\alpha$

$$\Rightarrow \quad \alpha = \sin^{-1}\left(\frac{r^{\frac{(n-1)}{2}}}{\sqrt{A}}\right) \qquad \ldots (6)$$

$$\Rightarrow \quad \int \frac{2\sqrt{A}\cos\alpha\, d\alpha}{(n-1)\sqrt{A}\cos\alpha\, d\alpha} = \int d\theta + c''$$

$$\Rightarrow \quad \alpha = \frac{(n-1)}{2}\theta + c''' \quad \text{(using (6))}$$

$$\Rightarrow \quad \sin^{-1}\left(\frac{r^{\frac{(n-1)}{2}}}{\sqrt{A}}\right) = \frac{(n-1)}{2}\theta + c'''.$$

Type V. To find the apsidal length of the orbit.

Algorithm to find the apsidal length of central orbit.

Step 1 Find h^2 using $h^2 = p^2v^2 = a^2{v_0}^2$, where a is apsidal distance and v_0 is velocity at apse.

Step 2 Put the value of f in $f = \frac{h^2}{p^3}\frac{dp}{dr} = \frac{\mu}{r^5}$ and integrate the result.

Step 3 Substitute the value of h^2 in the result obtained in step 2 and apply the apse condition $p = a$, hence evaluate the integration constant.

Step 4 Use $\frac{1}{p^2} = \frac{1}{r^2} + \frac{1}{r^4}\left(\frac{dr}{d\theta}\right)^2$ and find $\frac{dr}{d\theta}$.

Step 5 Use apse condition $\frac{dr}{d\theta} = 0$ and find the values of r which will be the required apsidal length.

ILLUSTRATIVE EXAMPLES

Example 1 A particle moves with a central acceleration $\mu\left(r - \frac{a^4}{r^3}\right)$ being projected from an apse at a distance a from the pole with velocity $2\sqrt{\mu}a$. Find the another apsidal length of the orbit.

Solution We have, according to question velocity at apse is $v_0 = 2\sqrt{\mu}a$

$$\Rightarrow \quad v_0^2 = 4\mu a^2 \qquad \dots (1)$$

Velocity at an Apse $v_0 = \dfrac{h}{a} \Rightarrow h^2 = a^2 v_0^2$

$$\Rightarrow \quad h^2 = a^2 4\mu a^2 \text{ (using (1))}$$

$$\Rightarrow \quad h^2 = 4\mu a^4 \qquad \dots (2)$$

We have $f = \dfrac{h^2}{p^3}\dfrac{dp}{dr} = \mu\left(r - \dfrac{a^4}{r^3}\right) \Rightarrow \dfrac{h^2}{p^3}dp = \mu\left(r - \dfrac{a^4}{r^3}\right)dr$

On integration we get

$$\int \frac{h^2}{p^3}dp = \int \mu\left(r - \frac{a^4}{r^3}\right)dr$$

$$\Rightarrow \quad -\frac{h^2}{2p^2} = \mu\left(\frac{r^2}{2} - \frac{a^4}{2r^2}\right) + c$$

$$\Rightarrow \quad \frac{h^2}{p^2} = \mu\left(\frac{a^4}{r^2} - r^2\right) + c' \text{ Using (2) we get,}$$

$$\Rightarrow \quad \frac{4\mu a^4}{p^2} = \mu\left(\frac{a^4}{r^2} - r^2\right) + c' \qquad \dots (3)$$

We know that at apse $\dfrac{dr}{d\theta} = 0$ and $p = r = a$ Using these condition in (4) we get,

$$\Rightarrow \quad \frac{4\mu a^4}{a^2} = \mu\left(\frac{a^4}{a^2} - a^2\right) + c' = 0 + c'$$

$$\Rightarrow \quad c' = 4\mu a^2 \qquad \dots (4)$$

In view of (3) and (4)

$$4\mu a^4\left(\frac{1}{r^2} + \frac{1}{r^4}\left(\frac{dr}{d\theta}\right)^2\right) = \mu\left(\frac{a^4}{r^2} - r^2\right) + 4\mu a^2$$

$$\Rightarrow \quad \frac{1}{r^4}\left(\frac{dr}{d\theta}\right)^2 = \left(\frac{1}{4r^2} - \frac{r^2}{4a^4}\right) + \frac{1}{a^2} - \frac{1}{r^2} = \frac{-r^4 + 4a^2r^2 - 3a^4}{4a^4r^2}$$

$$\Rightarrow \quad \left(\frac{dr}{d\theta}\right)^2 = \frac{r^2\left(-r^4+4a^2r^2-3a^4\right)}{4a^4} \qquad \ldots (5)$$

We know that at apse $\frac{dr}{d\theta}=0$. In view of (5) we get,

$$\frac{r^2\left(-r^4+4a^2r^2-3a^4\right)}{4a^4}=0$$

$$\Rightarrow \quad -r^4+4a^2r^2-3a^4=0$$

$$\Rightarrow \quad \left(a^2-r^2\right)\left(3a^2-r^2\right)=0$$

$$\Rightarrow \quad r=a \text{ and } r=a\sqrt{3}$$

Thus another apsidal distance is $a\sqrt{3}$

Example 4 A particle moves with a central acceleration $\mu\left(r^5-a^4r\right)$, being projected from an apse at a distance a from the pole with velocity $\sqrt{\frac{2\mu}{3}}a^3$. Find the another apsidal length of the orbit.

Solution We have, according to question velocity at apse is $v_0=\sqrt{\frac{2\mu}{3}}a^3$.

$$\Rightarrow \quad v_0^2=\frac{2\mu}{3}a^6 \qquad \ldots (1)$$

Velocity at an Apse $v_0=\frac{h}{a} \Rightarrow h^2=a^2v_0^2$

$$\Rightarrow \quad h^2=a^2\frac{2\mu}{3}a^6 \qquad \text{(using (1))}$$

$$\Rightarrow \quad h^2=\frac{2\mu}{3}a^8 \qquad \ldots (2)$$

We have $f=\frac{h^2}{p^3}\frac{dp}{dr}=\mu\left(r^5-a^4r\right) \Rightarrow \frac{h^2}{p^3}dp=\mu\left(r^5-a^4r\right)dr$

On integration we get

$$\int\frac{h^2}{p^3}dp=\int\mu\left(r^5-a^4r\right)dr$$

$$\Rightarrow \quad -\frac{h^2}{2p^2} = \mu\left(\frac{r^6}{6} - a^4\frac{r^2}{2}\right) + c$$

$$\Rightarrow \quad \frac{h^2}{p^2} = \mu\left(a^4r^2 - \frac{r^6}{3}\right) + c' \qquad \text{Using (2) we get,}$$

$$\Rightarrow \quad \frac{2\mu}{3}a^8\frac{1}{p^2} = \mu\left(a^4r^2 - \frac{r^6}{3}\right) + c' \qquad \ldots (3)$$

We know that at apse $\frac{dr}{d\theta} = 0$ and $p = r = a$. Using these conditions in (3) we get,

$$\Rightarrow \quad \frac{2\mu a^6}{3} = \frac{2\mu a^6}{3} + c' \Rightarrow c' = 0 \qquad \ldots (4)$$

In view of (3) and (4)

$$\frac{2\mu}{3}a^8\left(\frac{1}{r^2} + \frac{1}{r^4}\left(\frac{dr}{d\theta}\right)^2\right) = \mu\left(a^4r^2 - \frac{r^6}{3}\right)$$

$$\Rightarrow \quad \left(\frac{dr}{d\theta}\right)^2 = r^4\left(\frac{3r^2}{2a^4} - \frac{r^6}{2a^8} - \frac{1}{r^2}\right) = \frac{r^2}{2a^8}\left(3a^4r^4 - r^8 - 2a^8\right)$$

$$\Rightarrow \quad \left(\frac{dr}{d\theta}\right)^2 = \frac{r^2}{2a^8}\left(3a^4r^4 - r^8 - 2a^8\right) \qquad \ldots (5)$$

We know that at apse $\frac{dr}{d\theta} = 0$. In view of (5) we get,

$$\Rightarrow \quad 3a^4r^4 - r^8 - 2a^8 = 0$$

$$\Rightarrow \quad \left(r^4 - a^4\right)\left(r^4 - 2a^4\right) = 0$$

$$\Rightarrow \quad r = a \text{ and } r = \sqrt[4]{2}\,a$$

Thus another apsidal distance is $\sqrt[4]{2}\,a$.

REVIEW EXERCISE 3.2

1. Define an Apse in a central orbit. Show that at an Apse the particle moves perpendicular to radius vector.

2. A particle moves with a central acceleration $f=\frac{\mu}{r^3}$, if it is projected from an apse at a distance a from the origin with a velocity $=\frac{2}{a}\sqrt{\frac{\mu}{3}}$. Show that the equation of orbit is $r\cos\frac{\theta}{2}=a$.

3. A particle moves with a central acceleration $f=\mu\left(r+\frac{a^4}{r^3}\right)$, if it is projected from an apse at a distance a from the origin with a velocity $=2\sqrt{\mu a}$. Prove that its path is $r^2\left(2+\cos\sqrt{3}\theta\right)=3a^2$.

4. A particle moves with a central acceleration $f=\frac{\mu}{r^5}$, if it is projected from an apse at a distance a from the origin with a velocity $=\frac{\sqrt{\mu}}{a^2}$. Show that the equation of orbit is the circle $r=a$ or $x^2+y^2=a^2$.

5. A particle is projected from an apse at a distance a from the centre of force with a velocity $\frac{\sqrt{\mu}}{a}$ under the law of force

$$f=\mu\left\{\frac{2\left(a^2+b^2\right)}{r^5}-\frac{3a^2b^2}{r^7}\right\}.$$

Show that the orbit is $r^2=a^2\cos^2\theta+b^2\sin^2\theta$.

(**Hint.** for integration use the substitution $r^2=a^2\cos^2\alpha+b^2\sin^2\alpha$)

6. A particle is projected from an apse at a distance a from the centre of force with a velocity equal to that which it would be acquired from infinite under the effect of same central force $f=\frac{\mu}{r^n}$. Prove that the equation of orbit is $r^{\frac{(n-3)}{2}}=a^{\frac{(n-3)}{2}}\cos\frac{(n-3)}{2}\theta$.

7. A particle is projected from an apse at a distance a from the centre of force with a velocity equal to that which it would be acquired from infinite under the effect of same central force $f=\frac{\mu}{r^7}$. Prove that the equation of orbit is $r^2=a^2\cos 2\theta$.

8. A particle is projected from an apse at a distance a from the centre of force with a velocity equal to that which it would be acquired from infinite

under the effect of same central force $f = \frac{\mu}{r^5}$. Prove that the equation of orbit is the circle $r = a\cos\theta$ or $x^2 + y^2 = ax$.

9. If the law of force $f = \mu\left(\frac{1}{r^4} - \frac{10}{9}\frac{a}{r^5}\right)$ and the particle is projected from an apse at a distance $5a$ with velocity $= \sqrt{\frac{5}{7}}$ times that in a circle at the same distance. Show that the orbit is $r = a(3 + 2\cos\theta)$.

10. Show that the only law for a central attraction, for which the velocity in a circle at any distance is equal to the velocity acquired in falling from infinite to that distance by the same law of force, is that of the inverse cube. (**Hint.** $v_\infty^2 = 2\int_r^\infty f dr = v_0^2 = rf \Rightarrow -2\int_\infty^r f dr = rf$, now differentiate w.r. t. $rf + r\frac{df}{dr} = -2f \Rightarrow f = \frac{\mu}{r^3}$).

11. A particle is projected from an apse at a distance a from the centre of force with a velocity equal to $n\sqrt{\frac{\mu}{2a^4}}$ under the effect of the central force $f = \frac{\mu}{r^5}$. Prove that the other apsidal distance is $\frac{a}{\sqrt{n^2 - 1}}$.

12. A particle is projected from an apse at a distance a from the centre of force with a velocity equal to n times that which it would be acquired from infinite under the effect of same central force $f = \frac{\mu}{r^5}$. Prove that the other apsidal distance is $\frac{a}{\sqrt{n^2 - 1}}$.

13. A particle is projected from an apse at a distance a from the centre of force with a velocity $\frac{\sqrt{\mu}}{a}$ under the law of force

$f = \mu\left\{\frac{2(a^2 + b^2)}{r^5} - \frac{3a^2b^2}{r^7}\right\}$. Show that the other apsidal distance is b.

14. A particle is projected from an apse at a distance a from the centre of force with a velocity $2\sqrt{\mu a}$ under the law of force $f = \mu\left\{r + \frac{a^4}{r^3}\right\}$. Show that the other apsidal distance is $a\sqrt{3}$.

§ 6. Applications in Astronomy

§ 6.1. Newton's Law of Gravitation. *Every particle of the universe attracts every other particle with a force known as gravitational force. The gravitational force proportional to the product of their masses and inversely proportional to the square of the distance between them.* Thus for two bodies of masses m_1 and m_2 and distance between them is r, then *gravitational force* of attraction F

$F = \dfrac{Gm_1m_2}{r^2}$, where G is the gravitational constant.

Thus by virtue of Newton's Law of Gravitation, the law of force of the planetary motion is governed by inverse square law.

§ 6.2. Kepler's Laws of Planetary Motion. The German Astronomer **J. Kepler in 1619** stated the three laws by virtue of observations on the motions of planets of the solar system.

I. Each planet describes an ellipse about the sun and the sun is in one of its foci.

II. The radius vector drown from the sun to a planet sweeps out equal areas in equal times.

III. The squares of the periodic times of the various planets are proportional to the cubes of the semi major axis of their respective elliptical orbits.

Deduction of the Kepler's Laws of Planetary Motion from Newton's Law of Gravitation

I. Each planet describes an ellipse about the sun and the sun is in one of its foci. By virtue of Newton's Law of Gravitation, the law of force of the planetary motion is governed by inverse square law.

$$\Rightarrow \quad f = \frac{\mu}{r^2} = \mu u^2$$

$$\Rightarrow \quad \left(\frac{d^2u}{d\theta^2} + u\right) = \frac{f}{h^2u^2} = \frac{\mu u^2}{h^2u^2}$$

$$\Rightarrow \quad \left(\frac{d^2u}{d\theta^2} + u\right) = \frac{\mu}{h^2} \qquad \dots (1)$$

Multiplying both sides of (1) by $2\dfrac{du}{d\theta}$ and then integrate the result w.r. t. θ

$$2\frac{du}{d\theta}\frac{d^2u}{d\theta^2} + 2\frac{du}{d\theta}u = \frac{\mu}{h^2}2\frac{du}{d\theta}$$

$$\Rightarrow \quad \left(\frac{du}{d\theta}\right)^2 + u^2 = 2\frac{\mu}{h^2}u$$

$$\Rightarrow \quad \left(\frac{du}{d\theta}\right)^2 = -\left(u^2 - 2\frac{\mu}{h^2}u\right) = \frac{\mu^2}{h^4} - \left(u - \frac{\mu}{h^2}\right)^2$$

$$\Rightarrow \quad \int \frac{du}{\sqrt{\frac{\mu^2}{h^4} - \left(u - \frac{\mu}{h^2}\right)^2}} = \pm\int d\theta + C \quad \Rightarrow \quad \cos^{-1}\frac{\left(u - \frac{\mu}{h^2}\right)}{\frac{\mu}{h^2}} = \mp\theta + C$$

$$\Rightarrow \quad \left(u - \frac{\mu}{h^2}\right) = \frac{\mu}{h^2}A\cos(\theta - \alpha)$$

$$\Rightarrow \quad \frac{h^2/\mu}{r} - 1 = A\cos(\theta - \alpha) \quad \text{(Multiplying by } h^2/\mu \text{ and putting } u = \frac{1}{r}\text{)}$$

$$\Rightarrow \quad \frac{h^2/\mu}{r} = 1 + A\cos(\theta - \alpha)$$

$$\Rightarrow \quad \frac{l}{r} = 1 + A\cos(\theta - \alpha), \text{ where } l = h^2/\mu.$$

Thus the orbit is an ellipse, focus being the pole i.e. the position of the sun is at the focus.

II. The radius vector drown from the sun to a planet sweeps out equal areas in equal times. Let O be the centre of force as origin and affixed line

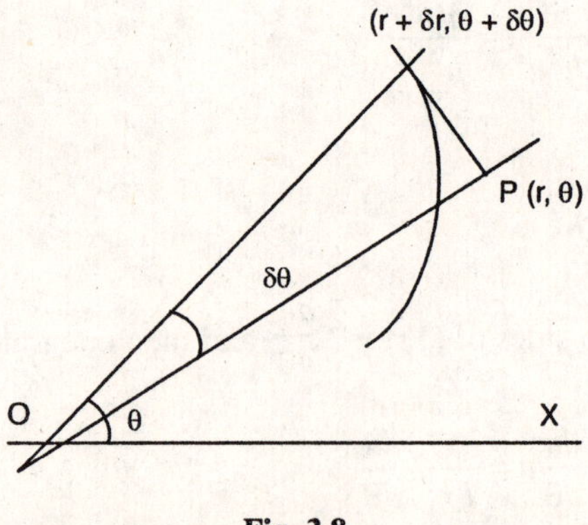

Fig. 3.8

OX as initial line let the particle be at point $P(r, \theta)$ at time t and after time $t+\delta t$ particle be at point $Q(r+\delta r, \theta+\delta\theta)$.

$$\text{Area } \Delta OPQ = \delta A = \frac{1}{2} OP OQ \sin\delta\theta$$

$$\Rightarrow \qquad \delta A = \frac{1}{2} r(r+\delta r)\sin\delta\theta$$

δA is the area swept during interval δt with the centre of force O by the particle.

Areal velocity = area swept in unit time

$$= \text{lt}_{\delta t\to 0}\frac{\delta A}{\delta t} = \text{lt}_{\delta t\to 0}\frac{\frac{1}{2}r(r+\delta r)\sin\delta\theta}{\delta t}$$

$$= \text{lt}_{\delta t\to 0}\frac{\frac{1}{2}r(r+\delta r)\delta\theta}{\delta t}$$

$$= \frac{1}{2}\left(r^2\dot{\theta}\right) = \frac{h}{2} \quad (\text{As} \sin\delta\theta \to \delta\theta \text{ for } \delta\theta \to 0 \text{ and } \delta r\delta\theta \to 0)$$

(For the motion of the planet around sun is governed by the central force therefore, **transverse component of acceleration is zero.** Transverse Acceleration $(f_t) = \frac{1}{r}\left(\frac{d\left(r^2\dot{\theta}\right)}{dt}\right) = 0$. $\Rightarrow$ $r^2\dot{\theta} = h$. Where h is a constant.)

$\Rightarrow$ Areal velocity $= \frac{h}{2}$, h is a constant.

Thus *area swept in unit time* $= \frac{h}{2}$, where h is a constant.

III. The squares of the periodic times of the various planets are proportional to the cubes of the semi major axis of their respective elliptical orbits

Area swept in unit time $= \frac{h}{2}$

Let T be the orbital time period i.e. time taken by the planet B P (r, θ) to complete one revolution.

$\Rightarrow$ Total area swept out in time T = Area of the ellipse

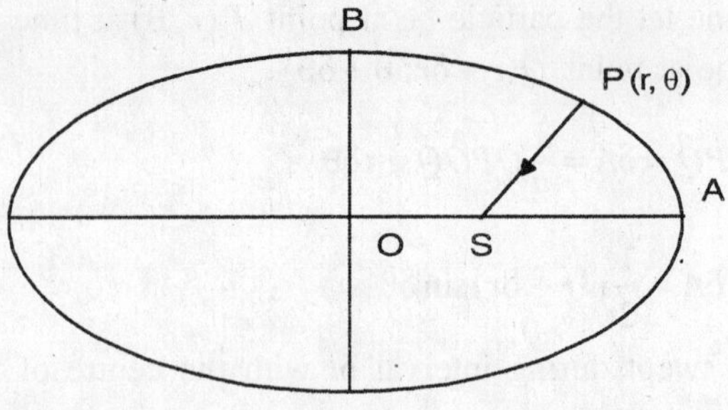

Fig. 3.9

$$\Rightarrow \quad \frac{h}{2} \times T = \pi ab$$

$$\Rightarrow \quad T = \frac{2\pi ab}{h} \qquad \ldots \text{(i)}$$

By Kepler's I Law we have, $h = \sqrt{\mu l}$, where $l = \frac{b^2}{a}$ = *semi latus rectum*, $a = OA$ = semi major axis and $b = OB$ = semi minor axis.

$$\Rightarrow \quad h = \sqrt{\mu \frac{b^2}{a}} \qquad \ldots \text{(ii)}$$

In view of (i) and (ii) we get, $T = \dfrac{2\pi ab}{\sqrt{\mu \frac{b^2}{a}}} = \dfrac{2\pi}{\sqrt{\mu}} a^{\frac{3}{2}}$

$$\Rightarrow \quad T^2 = \frac{4\pi^2}{\mu} a^3 \quad \Rightarrow \quad T^2 \propto a^3$$

Thus, the squares of the periodic times of the various planets are proportional to the cubes of the semi major axis (a).

Deduction of Newton's Law of Gravitation from the Kepler's Laws of Planetary Motion:

We have the equation of an ellipse $\frac{l}{r} = 1 + e\cos\theta$, where l is semi latus rectum and e is eccentricity of the ellipse.

$$\Rightarrow \quad u = \frac{1}{r} = \frac{1}{l} + \frac{e\cos\theta}{l} \qquad \ldots \text{(i)}$$

Differentiating twice (i) w.r. t. θ,we get,

$$\frac{du}{d\theta} = -\frac{e\sin\theta}{l}$$

$$\Rightarrow \quad \frac{d^2u}{d\theta^2} = -\frac{e\cos\theta}{l} \qquad \text{... (ii)}$$

Law of force of central orbit is given by

$$f = h^2u^2\left(\frac{d^2u}{d\theta^2} + u\right) \qquad \text{... (iii)}$$

In view of (i) and (ii), we get $f = h^2u^2\left(\frac{-e\cos\theta}{l} + \frac{1}{l} + \frac{e\cos\theta}{l}\right) = \frac{h^2u^2}{l}$

$$\Rightarrow \quad f = \frac{h^2u^2}{l} = \frac{h^2}{l}\frac{1}{r^2} \qquad \text{... (iv)}$$

From the **Kepler's III Law** $T^2 \propto a^3$

$$\Rightarrow \quad \frac{T^2}{a^3} = \text{constant} \qquad \text{... (v)}$$

The orbital time period is $T = \frac{\pi ab}{\frac{h}{2}} = \frac{2\pi ab}{h}$ ... (vi)

Substituting T from (vi) in (v), we get

$$\Rightarrow \quad \frac{T^2}{a^3} = \frac{4\pi^2b^2}{ah^2} = \frac{4\pi^2 l}{h^2} = \text{constant}$$

$$\Rightarrow \quad \frac{l}{h^2} = \text{constant, for all planets}$$

$$\Rightarrow \quad \frac{h^2}{l} = \text{constant} = \mu$$

Using (iv), we get

$$\Rightarrow \quad f = \frac{\mu}{r^2} \qquad \text{...(vii)}$$

Thus in view of (vii) we can say that in planetary motion the central force is inverse square of the distance. **Hence Newton's Law of Gravitation can be deduced from the Kepler's Laws of Planetary Motion.**

§ 6.3. Velocity of the planet. Since the orbit of the planet is specified as elliptical. Therefore, We have the equation of an ellipse $\frac{l}{r} = 1 + e\cos\theta$, where l is semi latus rectum and e is eccentricity of the ellipse.

$$\Rightarrow \quad u=\frac{1}{r}=\frac{1}{l}+\frac{e\cos\theta}{l} \qquad \text{... (i)}$$

Differentiating twice (i) w.r. t. θ, we get,

$$\frac{du}{d\theta}=-\frac{e\sin\theta}{l} \qquad \text{... (ii)}$$

$$\Rightarrow \quad \frac{d^2u}{d\theta^2}=-\frac{e\cos\theta}{l} \qquad \text{... (iii)}$$

Law of force of central orbit is given by $f=h^2u^2\left(\frac{d^2u}{d\theta^2}+u\right)$

In view of (i) and (iii), we get $f=h^2u^2\left(\frac{-e\cos\theta}{l}+\frac{1}{l}+\frac{e\cos\theta}{l}\right)=\frac{h^2u^2}{l}$

$$\Rightarrow \quad f=\frac{h^2u^2}{l}=\frac{h^2}{l}\frac{1}{r^2}=\frac{\mu}{r^2} \qquad \text{... (iv) where } \frac{h^2}{l}=\mu$$

We know that $v^2=\frac{h^2}{p^2}=h^2\left(u^2+\left(\frac{du}{d\theta}\right)^2\right)$

In view of (i), (ii) and $h^2=\mu l$

$$\Rightarrow \quad v^2=\mu l\left[\left(\frac{1}{l}+\frac{e\cos\theta}{l}\right)^2+\left(-\frac{e\sin\theta}{l}\right)^2\right]$$

$$=\mu l\left[\frac{e^2}{l^2}(\cos^2\theta+\sin^2\theta)+\frac{1}{l^2}+\frac{2e\cos\theta}{l^2}\right]$$

$$=\mu\left[\frac{e^2}{l}+\frac{1}{l}+\frac{2e\cos\theta}{l}\right]=\mu\left[2\left(\frac{1}{l}+\frac{e\cos\theta}{l}\right)-\frac{1-e^2}{l}\right]$$

$$=\mu\left[2u-\frac{b^2/a^2}{b^2/a}\right] \qquad \left(\text{As } \frac{b^2}{a^2}=1-e^2 \text{ and } l=\frac{b^2}{a}\right)$$

$\Rightarrow v^2=\mu\left[\frac{2}{r}-\frac{1}{a}\right]$, here a = semi major axis, b = semi minor axis and $u=\frac{1}{r}$.

Important results to be used

1 The periodic time $T = \frac{2\pi}{\sqrt{\mu}} a^{\frac{3}{2}}$,

2. Velocity of the planet v then $v^2 = \mu\left[\frac{2}{r} - \frac{1}{a}\right]$

3. $h^2 = \mu l$ where $l = \frac{b^2}{a}$, a = semi major axis, b = semi minor axis

4. For the nearest point of planet from sun $r = (a - ae)$

5. For the farthest point of planet from sun $r = (a + ae)$, eccentricity

$e = \sqrt{1 - \frac{b^2}{a^2}}$

Example 1 A particle describes an ellipse under a force $\frac{\mu}{r^2}$, towards the focus. If it was projected with velocity V from a point at a distance d from the centre of force, show that its periodic time is $\frac{2\pi}{\sqrt{\mu}}\left(\frac{2}{d} - \frac{V^2}{\mu}\right)^{-3/2}$

Solution For an elliptical orbit, we have

$$\Rightarrow \quad v^2 = \mu\left[\frac{2}{r} - \frac{1}{a}\right]$$

According to question initially, velocity $v = V$ when $r = d$.

$$\Rightarrow \quad V^2 = \mu\left[\frac{2}{d} - \frac{1}{a}\right]$$

$$\Rightarrow \quad \frac{V^2}{\mu} - \frac{2}{d} = -\frac{1}{a}$$

$$\Rightarrow \quad a = \frac{1}{\left(\frac{2}{d} - \frac{V^2}{\mu}\right)} \qquad \dots (1)$$

The periodic time T is given by

$$T = \frac{2\pi}{\sqrt{\mu}} a^{\frac{3}{2}} \qquad \text{(using (1))}$$

$$\Rightarrow \quad T = \frac{2\pi}{\sqrt{\mu}} \left(\frac{2}{d} - \frac{V^2}{\mu} \right)^{-\frac{3}{2}}$$

Example 2 If v_1 and v_2 are the velocities of a planet when it is nearest and farthest points respectively from the sun, show that $(1-e)v_1 = (1+e)v_2$.

Solution For an elliptical orbit, we have

$$\Rightarrow \quad v^2 = \mu \left[\frac{2}{r} - \frac{1}{a} \right]$$

According to question, velocity $v = v_1$ when $r = (a - ae)$ and velocity $v = v_2$ when $r = (a + ae)$ the sun is at focus S i.e. $S(ae, 0)$.

$$\Rightarrow \quad v_1^2 = \mu \left[\frac{2}{a - ae} - \frac{1}{a} \right] = \frac{\mu}{a} \frac{(1+e)}{(1-e)} \qquad \dots (1)$$

$$v_2^2 = \mu \left[\frac{2}{a + ae} - \frac{1}{a} \right] = \frac{\mu}{a} \frac{(1-e)}{(1+e)} \qquad \dots (2)$$

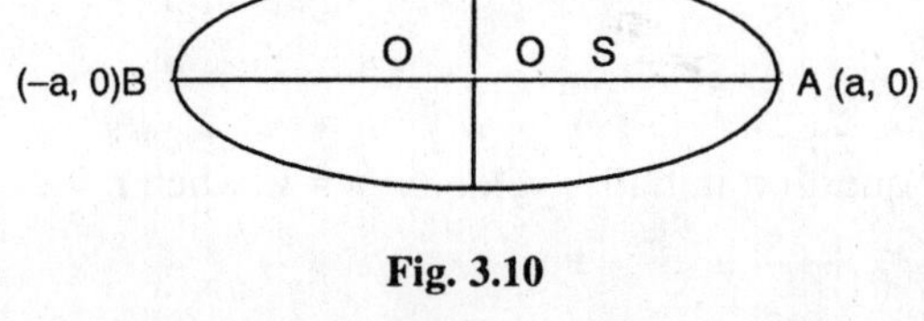

Fig. 3.10

In view of (1) and (2) we get, $\dfrac{v_1^2}{v_2^2} = \left(\dfrac{1+e}{1-e} \right)^2$

$$\Rightarrow \quad \frac{v_1}{v_2} = \left(\frac{1+e}{1-e} \right) \Rightarrow (1-e)v_1 = (1+e)v_2.$$

Example 3 If v_1 and v_2 are the velocities of a planet when it is nearest and farthest points respectively from the sun and v_3 is the velocity at the end of minor axis, show that $v_3^2 = v_1 v_2$.

Solution For an elliptical orbit, we have

$$\Rightarrow \quad v^2 = \mu\left[\frac{2}{r} - \frac{1}{a}\right]$$

According to question, velocity $v = v_1$ when $r = (a - ae)$ and velocity $v =$ v_2 when $r = (a + ae)$ the sun is at focus S i.e. S.

$$\Rightarrow \quad v_1^2 = \mu\left[\frac{2}{a - ae} - \frac{1}{a}\right] = \frac{\mu}{a}\frac{(1+e)}{(1-e)} \qquad \ldots (1)$$

$$v_2^2 = \mu\left[\frac{2}{a + ae} - \frac{1}{a}\right] = \frac{\mu}{a}\frac{(1-e)}{(1+e)} \qquad \ldots (2)$$

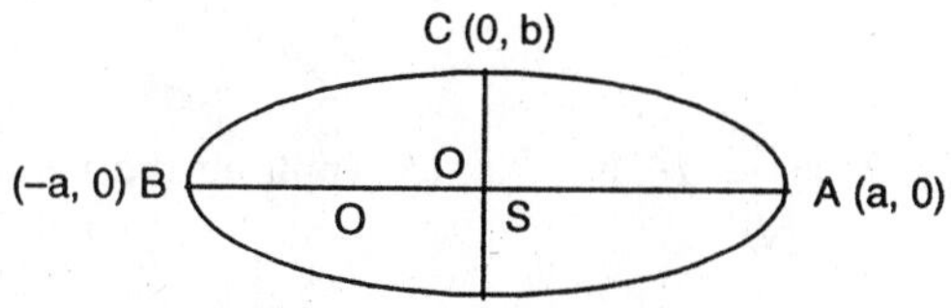

Fig. 3.11

For point C (0, b) the distance is r from S (ae, 0)

$$\Rightarrow r = \sqrt{(0 - ae)^2 + (b - 0)^2} = \sqrt{(ae)^2 + b^2} = a \text{ (put the value of } e = \sqrt{1 - \frac{b^2}{a^2}}\text{)}$$

$$\Rightarrow \quad v_3^2 = \mu\left[\frac{2}{a} - \frac{1}{a}\right] = \frac{\mu}{a} \qquad \ldots (3)$$

In view of (1) and (2) we get, $v_1^2 v_2^2 = \frac{\mu}{a}\frac{(1-e)}{(1+e)}\frac{\mu}{a}\frac{(1+e)}{(1-e)} = \left(\frac{\mu}{a}\right)^2$ (using (3))

$$\Rightarrow \quad v_1^2\, v_2^2 = v_3^4$$

$$\Rightarrow \quad v_3^2 = v_1 v_2.$$

Example 3 If ω is the velocity of a planet when it is nearest to the sun, prove that its period is $T = \frac{2\pi}{\omega}\sqrt{\frac{(1+e)}{(1-e)^3}}$.

Solution According to question, angular velocity = ω when $r = (a - ae)$, the nearest point to the sun and the sun is at focus S i.e. S. (ae, 0).

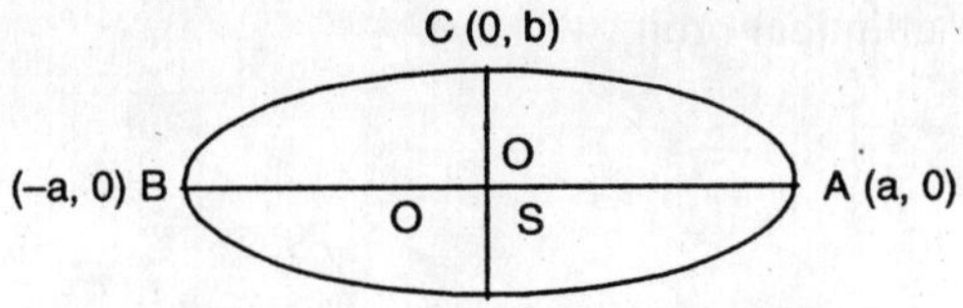

Fig. 3.12

For central orbit $r^2\dot{\theta} = h = \sqrt{\mu l}$

$$\Rightarrow \quad (a - ae)^2 \omega = \sqrt{\mu a(1 - e^2)} \qquad \dots (1)$$

$$\Rightarrow \quad (a - ae)^4 \omega^2 = \mu a(1 - e^2)$$

$$\Rightarrow \quad \mu = \frac{a^3(1-e)^3}{1+e}\omega^2 \qquad \dots (2)$$

The periodic time $T = \frac{2\pi}{\sqrt{\mu}} a^{\frac{3}{2}}$ ... (3) (putting the value of μ from (2))

$$\Rightarrow \quad T = \frac{2\pi}{\omega}\sqrt{\frac{(1+e)}{(1-e)^3}}.$$

Example 4 Show that the velocity of a planet moving in an ellipse about the centre of force at the focus is compounded of two constant velocities $\frac{\mu}{h}$, is perpendicular to radius vector and $e\frac{\mu}{h}$ is perpendicular to the major axis.

Solution Let the equation of the ellipse be

$$\frac{l}{r} = 1 + e\cos\theta \qquad \dots (1)$$

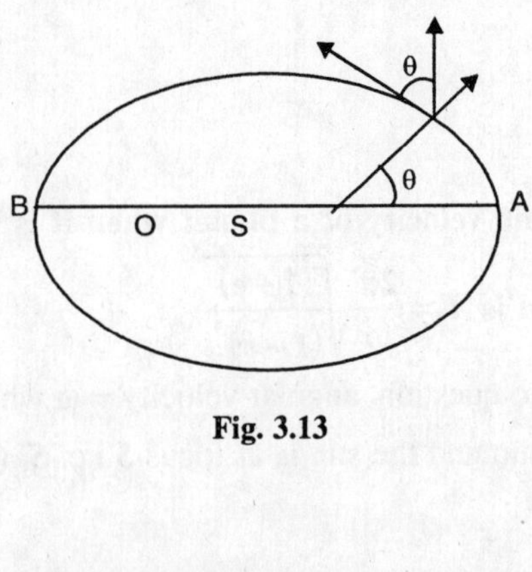

Fig. 3.13

We know that for central force

$$r^2\dot{\theta} = h, \qquad l = \frac{h^2}{\mu} \qquad \dots (2)$$

We know that the radial velocity $v_r = \dot{r}$

Differentiating (1) w.r. t. t, we get, $-\frac{l}{r^2}\dot{r} = -e\sin\theta\dot{\theta}$

$$\Rightarrow \quad v_R = \dot{r} = \frac{e\sin\theta r^2\dot{\theta}}{l} = \frac{eh\sin\theta}{l} \qquad \text{(using (2))}$$

$$\Rightarrow \quad v_r = \frac{e\mu\sin\theta}{h} \qquad \dots (3)$$

Similarly We know that the transverse velocity $v_t = r\dot{\theta}$

$$\Rightarrow \quad v_t = r\dot{\theta} = \frac{rh}{r^2} = \frac{h}{r} \qquad \text{(using (2))}$$

$$\Rightarrow \quad v_t = h\frac{1+e\cos\theta}{l} = h\frac{1+e\cos\theta}{\frac{h^2}{\mu}} \qquad \text{(using (1) and (2))}$$

$$\Rightarrow \quad v_t = \mu\frac{1+e\cos\theta}{h} = \frac{\mu}{h} + \frac{e\mu}{h}\cos\theta \qquad \dots (4)$$

In view of (3) and (4) we can say that the velocity of the planet is the composition of three velocities:

(i) $\frac{\mu}{h}$ in the transverse direction i.e. perpendicular to radius vector... (5)

(ii) $\frac{e\mu}{h}\cos\theta$ in the transverse direction ... (6)

(iii) $\frac{e\mu\sin\theta}{h}$ in the radial direction ... (7)

The resultant of (ii) and (iii) $= \sqrt{\left(\frac{e\mu}{h}\cos\theta\right)^2 + \left(\frac{e\mu}{h}\sin\theta\right)^2} = \frac{e\mu}{h}$

Angle between resultant of (ii) and (iii) with transverse direction ϕ, we have

$$\Rightarrow \quad \phi = \tan^{-1}\left(\frac{\dfrac{e\mu \sin\theta}{h}}{\dfrac{e\mu}{h}\cos\theta}\right) = \tan^{-1}\tan\theta = \theta$$

Thus the angle between resultant of (ii) and (iii) with radial direction $= (90° - \theta)$ = perpendicular to the major axis (see Fig.).

$$\Rightarrow \quad \frac{e\,\mu}{h} \text{ is the component of velocity perpendicular to major axis} \qquad \dots (8)$$

In view of (5) and (8) we get the required result.

4 Hodograph

§ 1. Definition

If a particle describes a curve $r = f(\theta)$, then corresponding to each point on the curve, the particle possesses the definite velocity. If we trace the curve corresponding to these values of velocities assuming the same origin as $r = f(\theta)$. Let the velocity curve be $R = F(\theta)$, then $R = F(\theta)$ is known as *Hodograph of* $r = f(\theta)$. In other words *we can say that the hodograph of a moving point P whose position vector is* $\vec{r}$ *of* $r = f(\theta)$ *is the locus of another point M such that the position vector* $\vec{R}$ of *M is equal to velocity vector* $\vec{v}$ *of P. In figure* C_2 *is known as hodograph of* C_1, *where* $\vec{v} = \vec{R}$.

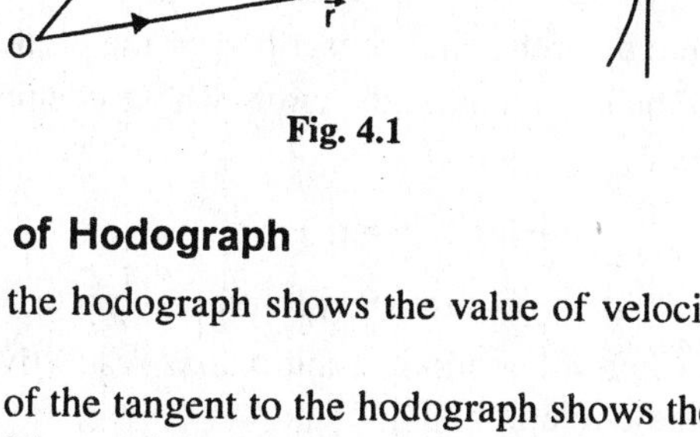

Fig. 4.1

Characteristics of Hodograph

1. Any point on the hodograph shows the value of velocity of the moving particle.
2. The direction of the tangent to the hodograph shows the direction of the acceleration of the particle at that instant i.e. the direction of tangent at *M* is the direction of acceleration at *P*.

3. The velocity at any point M of the hodograph is equal to the acceleration of the particle at the corresponding point P on the curve.

Applications of Hodograph

1. *If the equation of hodograph is known then with the help of equation of hodograph, we can find the velocity at each point on the path of the particle.*
2. *The direction of acceleration of the particle can be obtained, as direction of the tangent to the hodograph shows the direction of the acceleration of the particle at that instant.*
3. ***The acceleration of the particle can be obtained,** as the acceleration of the particle at point P on the curve is equal to the velocity at corresponding point M of the hodograph.*

§ 2. Algorithm to find the Hodograph. Algorithm to find the Hodograph involves the following steps:

Step 1. Write the equation of curve whose hodograph is required.

Step 2. Find the velocity vector $\vec{v}$ by differentiating equation in step 1 with respect to time (t).

Step 3. Simplify equation for $\vec{v}$ obtained in step 2 with the help of given condition, if any.

Step 4. Replace $\vec{v}$ by $\vec{R}$ in equation obtained in step 3.

Step 5. Put $\vec{R} = xi + yj$ in step 4 and simplify it.

Note. *Step 4 gives the equation of hodograph in polar coordinates, whereas step 5 gives the equation of hodograph in Cartesian coordinates.*

ILLUSTRATIVE EXAMPLES

Example 1 A particle describes a circle with uniform speed. Show that the hodograph is a circle of constant speed.

Solution Let the radius of the circle described by the particle be a with centre O, as origin. Let P be any point on the circle whose position vector be $\vec{r}$ and velocity vector be $\vec{v}$.

$$|\vec{r}| = a,\ \vec{r} = xi + yj, \qquad \dots (1)$$

$$x = a\cos\theta,\ y = a\sin\theta, \qquad \dots (2)$$

Where i and j are unit vector along x and y axis respectively.

The equation of the circle is

$$\vec{r} = a\cos\theta i + a\sin\theta j \qquad \dots (3)$$

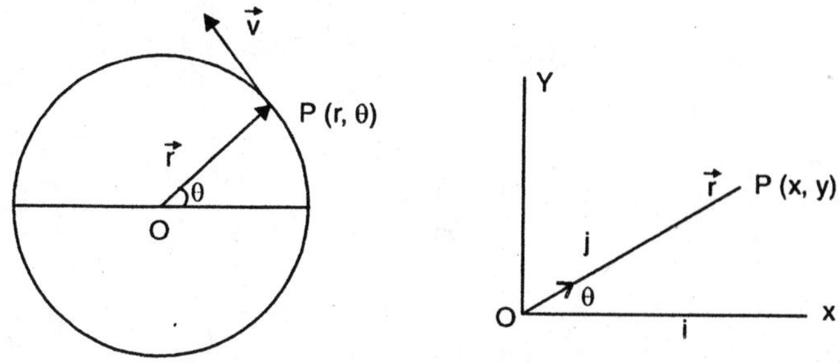

Fig. 4.2

Differentiating (3) w.r. t. t we get,

$$\frac{d\vec{r}}{dt} = (-a\sin\theta i + a\cos\theta j)\frac{d\theta}{dt}$$

$$\Rightarrow \quad \vec{v} = (-a\sin\theta i + a\cos\theta j)\frac{d\theta}{dt} \qquad \dots (4)$$

According to question speed of the particle is constant

$$\Rightarrow \quad |\vec{v}| = \left|(-a\sin\theta i + a\cos\theta j)\frac{d\theta}{dt}\right| = k$$

$$\Rightarrow \quad |\vec{v}| = \sqrt{\left(-a\sin\theta\frac{d\theta}{dt}\right)^2 + \left(a\cos\theta\frac{d\theta}{dt}\right)^2} = \left(a\frac{d\theta}{dt}\right) = k$$

$$\Rightarrow \quad \left(a\frac{d\theta}{dt}\right) = k \qquad \dots (5)$$

For the hodograph, take the locus of (4) i.e. replacing $\vec{v}$ with $\vec{R}$ we get,

$$\vec{R} = (-a\sin\theta i + a\cos\theta j)\frac{d\theta}{dt} \qquad \text{(using (5))}$$

$$\Rightarrow \quad \vec{R} = k(-\sin\theta i + \cos\theta j) \qquad \dots (6)$$

If $\vec{R} = xi + yj$, $x = -k\sin\theta$ and $y = k\cos\theta$

$$\Rightarrow \quad x^2 + y^2 = k^2 \qquad \dots (7)$$

The hodograph is represented by (6) and (7) in polar and Cartesian form respectively. Thus hodograph is a circle.

Differentiating (6) w.r. t. t we get,

$$\frac{d\vec{R}}{dt} = k(-\cos\theta i - \sin\theta j)\frac{d\theta}{dt} \qquad \text{(using (5))}$$

$$\Rightarrow \quad \vec{V} = k(-\cos\theta i - \sin\theta j)\frac{k}{a}$$

$$\Rightarrow \quad |\vec{V}| = \left|\frac{k^2}{a}(-\cos\theta i - \sin\theta j)\right| = \sqrt{\left(-\frac{k^2}{a}\cos\theta\right)^2 + \left(-\frac{k^2}{a}\sin\theta\right)^2}$$

$$\Rightarrow \quad |\vec{V}| = \frac{k^2}{a} \text{constant.}$$

Thus hodograph is a circle of constant speed or the acceleration of the particle is constant.

Example 2 A particle is moving such that its hodograph is a circle having constant angular speed about a point O on the circumference. Show that the path of the particle is a cycloid.

Solution Let the radius of the hodograph circle be a and O as origin on the circumference. Let P be any point on the circle whose position vector be $\vec{R}$ = velocity vector $\vec{v}$ of the particle.

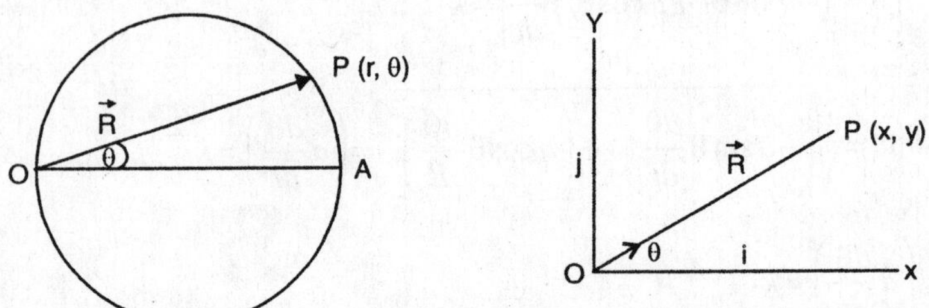

Fig. 4.3

$$|\vec{R}| = 2a\cos\theta, \ \vec{R} = xi + yj \qquad \dots (1)$$

$$x = 2a\cos\theta\cos\theta, \ y = 2a\cos\theta\sin\theta \qquad \dots (2)$$

Where i and j are unit vector along x and y axis respectively.
The equation of the hodograph circle is

$$\vec{R} = 2a\cos\theta\cos\theta i + 2a\cos\theta\sin\theta j \qquad \dots (3)$$

We know that position vector of hodograph $\vec{R}$ = velocity vector $\vec{v}$ of the particle

$$\Rightarrow \quad \vec{v} = 2a\cos\theta\cos\theta i + 2a\cos\theta\sin\theta j,$$

(using $\vec{v} = \frac{d\vec{r}}{dt}$, $\vec{r}$ is the position vector of any point of the path)

$$\Rightarrow \quad \frac{d\vec{r}}{dt} = 2a\cos^2\theta i + 2a\cos\theta\sin\theta j \qquad \ldots (4)$$

According to question hodograph is a circle having constant angular speed

$$\Rightarrow \quad \text{Angular velocity} = \frac{d\theta}{dt} = k, \text{ where } k \text{ is a constant.}$$

$$\Rightarrow \quad \frac{1}{k}\frac{d\theta}{dt} = 1 \qquad \ldots (5)$$

In view of (5), (4) reduces to

$$\frac{d\vec{r}}{dt} = \left(2a\cos^2\theta i + 2a\cos\theta\sin\theta j\right)\frac{1}{k}\frac{d\theta}{dt}$$

$$\Rightarrow \quad d\vec{r} = \left(2a\cos^2\theta i + 2a\cos\theta\sin\theta j\right)\frac{1}{k}d\theta \qquad \ldots (6)$$

(using $2\cos^2\theta = 1 + \cos 2\theta$, $2\cos\theta\sin\theta = \sin 2\theta$)

Now on integration (6) we get,

$$\vec{r} = \frac{a}{k}\left\{\left(\theta + \frac{\sin 2\theta}{2}\right)i - \left(\frac{\cos 2\theta}{2}\right)j\right\} + C \qquad \ldots (7)$$

When $\theta = 0 \Rightarrow \vec{r} = bi \Rightarrow$ as the vector is along x-axis only. Applying the conditions in (7) we get,

$$bi = -\frac{a}{2k}j + C \quad \Rightarrow \quad C = bi + \frac{a}{2k}j$$

Now putting the values of C and $\vec{r} = xi + yj$ in (7) we get,

$$xi + yj = \frac{a}{k}\left\{\left(\theta + \frac{\sin 2\theta}{2}\right)i - \left(\frac{\cos 2\theta}{2}\right)j\right\} + bi + \frac{a}{2k}j$$

$$\Rightarrow \quad \boldsymbol{x - b = \frac{a}{2k}(2\theta + \sin 2\theta) \text{ and } y = \frac{a}{2k}(1 - \cos 2\theta)} \qquad \ldots (8)$$

Equation (8) is the parametric representation of cycloid.

§ 3. Algorithm to Find the Hodograph of a central orbit. Algorithm to find the Hodograph **of central orbit** involves the following steps:

Step 1 Write the equation of force whose hodograph is required i.e. $\frac{d^2\vec{r}}{dt^2} = \frac{d\vec{v}}{dt} = f(r)\,\hat{r} = f(r)(\cos\theta i + \sin\theta j)$ where $\hat{r} = \cos\theta i + \sin\theta j$, is a unit vector i.e. $|\hat{r}| = \sqrt{\cos^2\theta + \sin^2\theta} = 1$.

Step 2 For central orbit $r^2\frac{d\theta}{dt}=h \Rightarrow \frac{r^2}{h}\frac{d\theta}{dt}=1.$

Step 3 Use step 2 in step-1 i.e. $\frac{d\vec{v}}{dt}=f(r)\hat{r}\times 1=f(r)\hat{r}\frac{r^2}{h}\frac{d\theta}{dt}$

$$\Rightarrow \quad d\vec{v}=\frac{r^2}{h}f(r)(\cos\theta i+\sin\theta j)\,d\theta.$$

Step 4. Now put the value of r from the equation of the orbit, if it is in equation of step 3.

Step 5. Find the velocity vector $\vec{v}$ by integrating equation of step 4.

Step 6. Replace $\vec{v}$ by $\vec{R}$ in equation obtained in step 5.

Step 7. Put $\vec{R}=xi+yj$ in step 4 and simplify it.

Note. *Step 6 gives the equation of hodograph in polar coordinates, whereas step 7 gives the equation of hodograph in Cartesian coordinates.*

Example 3 A particle describes a path under the central attraction force varies inversely as the square of the distance from a fixed point, show that the hodograph is a circle.

Solution According to question the given central force is inverse square law therefore, we have

$$\frac{d^2\vec{r}}{dt^2}=-\frac{\mu}{r^2}\hat{r} \qquad \dots (1)$$

(**Note.** *negative sign shows that the force is attraction in nature*)

$$\Rightarrow \quad \frac{d^2\vec{r}}{dt^2}=\frac{d\vec{v}}{dt}=-\frac{\mu}{r^2}\hat{r}=-\frac{\mu}{r^2}(\cos\theta i+\sin\theta j),$$

$$\Rightarrow \quad \frac{d\vec{v}}{dt}=-\frac{\mu}{r^2}(\cos\theta i+\sin\theta j)\frac{r^2}{h}\frac{d\theta}{dt}, \text{ (See step 2 of algorithm)}$$

$$\Rightarrow \quad d\vec{v}=-\frac{\mu}{h}(\cos\theta i+\sin\theta j)\,d\theta \qquad \dots (2)$$

Now on integration (2) we get,

$$\vec{v}=-\frac{\mu}{h}(sin\theta i-cos\theta j)+C \qquad \dots (3)$$

For the hodograph, take the locus of (3) i.e. replacing $\vec{v}$ with $\vec{R}$ we get,

$$\vec{R} = -\frac{\mu}{h}(\sin\theta i - \cos\theta j) + C \qquad \dots (4)$$

If $\vec{R} = xi + yj$ and $C = h_1 i + k_1 j$, put in (4) we get,

$$\Rightarrow \quad x - h_1 = -\frac{\mu}{h}\sin\theta \text{ and } y - k_1 = \frac{\mu}{h}\cos\theta$$

$$\Rightarrow \quad (x - h_1)^2 + (y - k_1)^2 = \left(\frac{\mu}{h}\right)^2 \qquad \dots (5)$$

Which is the equation of a circle of radius $\frac{\mu}{h}$.

Example 4 A particle describes an equiangular spiral $r = ae^{c\theta}$ under the effect of central force $f = \frac{\mu}{r^3}$ directed to the pole. Prove that the hodograph is also an equiangular spiral.

Solution According to question, the given central force is attraction and $f = \frac{\mu}{r^3}$, therefore, we have,

$$\frac{d^2\vec{r}}{dt^2} = -\frac{\mu}{r^3}\hat{r} \qquad \dots (1)$$

(**Note.** *Negative sign shows that the force is attraction i.e. directed to the pole*)

$$\Rightarrow \quad \frac{d^2\vec{r}}{dt^2} = \frac{d\vec{v}}{dt} = -\frac{\mu}{r^3}\hat{r} = \frac{\mu}{r^3}(\cos\theta i + \sin\theta j),$$

$$\Rightarrow \quad \frac{d\vec{v}}{dt} = -\frac{\mu}{r^3}(\cos\theta i + \sin\theta j)\frac{r^2}{h}\frac{d\theta}{dt}, \text{ (See step 2 of algorithm)}$$

$$\Rightarrow \quad d\vec{v} = -\frac{\mu}{hr}(\cos\theta i + \sin\theta j)\,d\theta \text{ (put } r = ae^{c\theta}\text{)}$$

$$\Rightarrow \quad d\vec{v} = -\frac{\mu}{ha}(\cos\theta i + \sin\theta j)e^{-c\theta}d\theta \qquad \dots (2)$$

Now on integration (2) we get,

$$\vec{v} = -\frac{\mu}{ha}\frac{e^{-c\theta}}{(1+c^2)}\{(-c\cos\theta + \sin\theta)i + (-c\sin\theta - \cos\theta)\} + D \qquad \dots (3)$$

For the hodograph, take the locus of (3) i.e. replacing $\vec{v}$ with $\vec{R}$ we get,

$$\vec{R} = -\frac{\mu}{ha}\frac{e^{-c\theta}}{(1+c^2)}\{(-c\cos\theta + \sin\theta)i + (-c\sin\theta - \cos\theta)j\} + D \quad \ldots (4)$$

If $\vec{R} = xi + yj$ and $D = h_1 i + k_1 j$, put in (4) we get,

$$\Rightarrow \quad x - h_1 = -\frac{\mu}{ha}\frac{e^{-c\theta}}{(1+c^2)}(-c\cos\theta + \sin\theta)$$

and $y - k = -\dfrac{\mu}{ha}\dfrac{e^{-c\theta}}{(1+c^2)}(-csin\theta - cos\theta)$

$$\Rightarrow \quad (x-h_1)^2 + (y-k_1)^2 = \left(-\frac{\mu}{ha}\frac{e^{-c\theta}}{(1+c^2)}\right)^2$$

$$\left\{(-c\cos\theta + \sin\theta)^2 + (-c\sin\theta - \cos\theta)^2\right\} \ldots (5)$$

$$\Rightarrow \quad (x-h_1)^2 + (y-k_1)^2 = \left(-\frac{\mu}{ha}\frac{e^{-c\theta}}{(1+c^2)}\right)^2 (1+c^2)$$

$$\Rightarrow \quad (x-h_1)^2 + (y-k_1)^2 = \frac{\mu^2}{h^2a^2(1+c^2)}e^{-2c\theta} \quad \ldots (6)$$

Now converting Cartesian coordinates into polar coordinates after shifting the origin i.e. $x - h_1 = X = r\cos\theta$, $y - k_1 = Y = r\sin\theta$, and putting in (6) we get, $r^2 = b^2 e^{-2c\theta}$, where $b^2 = \dfrac{\mu^2}{h^2a^2(1+c^2)} \Rightarrow r = be^{-c\theta}$, which is the equation of equiangular spiral.

Example 5 A particle describes a circle $r = 2a\cos\theta$ under the effect of central force $f = \dfrac{\mu}{r^5}$ directed to the point on the circumference. Prove that the hodograph is a parabola.

Solution According to question the given central force is attraction and $f = \dfrac{\mu}{r^5}$, therefore, we have,

$$\frac{d^2\bar{r}}{dt^2} = -\frac{\mu}{r^5}\hat{r} \qquad \text{... (1)}$$

(**Note.** *negative sign shows that the force is attraction i.e. directed to the pole*)

$$\Rightarrow \quad \frac{d^2\bar{r}}{dt^2} = \frac{d\bar{v}}{dt} = -\frac{\mu}{r^5}\hat{r} = \frac{\mu}{r^5}(\cos\theta i + \sin\theta j)$$

$$\Rightarrow \quad \frac{d\bar{v}}{dt} = -\frac{\mu}{r^5}\cos\theta i + \sin\theta j \frac{r^2}{h}\frac{d\theta}{dt}, \text{ (See step 2 of algorithm)}$$

$$\Rightarrow \quad d\bar{v} = -\frac{\mu}{hr^3}(\cos\theta i + \sin\theta j)\,d\theta \text{ (put } r = 2a\cos\theta\text{)}$$

$$\Rightarrow \quad d\bar{v} = -\frac{\mu}{8ha^3\cos^3\theta}(\cos\theta i + \sin\theta j)\,d\theta$$

$$\Rightarrow \quad d\bar{v} = -\frac{\mu}{8ha^3}(\sec^2\theta i + \sec^2\theta\tan\theta j)\,d\theta \qquad \text{... (2)}$$

Now on integration (2) we get,

$$\bar{v} = -\frac{\mu}{8ha^3}\left(\tan\theta i + \frac{\tan^2\theta}{2}j\right) + D \qquad \text{... (3)}$$

For the hodograph, take the locus of (3) i. e. replacing $\bar{v}$ and $\vec{R}$ we get,

$$\vec{R} = -\frac{\mu}{8ha^3}\left(\tan\theta i + \frac{\tan^2\theta}{2}j\right) + D \qquad \text{... (4)}$$

If $\vec{R} = xi + yj$ and $D = h_1 i + k_1 j$, put in (4) we get,

$$\Rightarrow \quad x - h_1 = -\frac{\mu}{8ha^3}\tan\theta$$

$$\Rightarrow \quad \tan\theta = \frac{x-h}{-\frac{\mu}{8ha^3}} = \frac{8ha^3(x-h_1)}{\mu} \qquad \text{... (5)}$$

$$y - k_1 = -\frac{\mu}{8ha^3}\frac{\tan^2\theta}{2} \qquad \text{... (6)}$$

Putting the value of $\tan\theta$ from (5) in (6) we get,

$$y-k_1=-\frac{\mu}{8ha^3}\frac{\left(\frac{8ha^3(x-h_1)}{\mu}\right)^2}{2}$$

$$\Rightarrow \quad (x-h_1)^2=-\frac{\mu}{4ha^3}(y-k_1) \qquad \ldots (7)$$

Which is the equation of a parabola of the type $X^2=-4bY$, with vertex (h_1,k_1).

REVIEW EXERCISE 4.1

1. A particle describes a circle $r = a$ with uniform speed. show that the hodograph is a circle.
2. A particle describes a conic under the central force to the focus, show that the hodograph is a circle. (**Hint.** A particle describes a conic under the central force to the focus implies the central attraction force varies inversely as the square of the distance from a fixed point)
3. A particle describes a circle under the effect of central force $f=\frac{\mu}{r^5}$ directed to the point on the circumference. Prove that the hodograph is a parabola. (**Hint.** Equation of circle with circumference as pole is $r=2a\cos\theta$)
4. A particle describes an ellipse under the central force $f=\mu r$ directed to its centre, show that the hodograph is an ellipse.
5. Prove that the hodograph of the path of projectile is a vertical line described with uniform velocity. (**Hint.** For projectile $\dot{x}=u\cos\alpha=\text{constant}$ and $\dot{y}=u\cos\alpha-gt \Rightarrow$ for hodograph $x=u\cos\alpha=\text{constant}$ is a line parallel to y–axis and $y=u\cos\alpha-gt$, velocity of hodograph $=\frac{dy}{dt}=-g=$ constant.)

5 Simple Harmonic Motion (SHM)

Definition

A particle is said to execute Simple harmonic motion if it moves in a straight line such that its acceleration is always directed towards a fixed point and proportional to the distance of the particle from the fixed point.

§ 1. *Acceleration, Velocity and Position of particle (SHM) at any instant.*

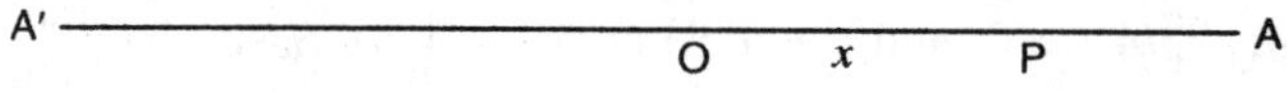

Fig. 5.1

Let O be the fixed point and let the particle start at rest from point A, such that $OA = a$. Let after time t particle be at P, such that $OP = x$.

The equation of motion of the particle is given by

$$\frac{d^2x}{dt^2} = -\mu x \qquad \text{(i)}$$

Acceleration of the particle executing SHM at any instant (–ve sign indicates that x decreases with the increase of time t), where μ is the constant of proportionality.

Multiplying both sides of (i) by $2\frac{dx}{dt}$, we get $2\frac{dx}{dt}\frac{d^2x}{dt^2} = -\mu x 2\frac{dx}{dt}$ and on integration we get

$$\left(\frac{dx}{dt}\right)^2 = -\mu x^2 + C \qquad \text{(ii)}$$

C is the constant of integration.

At point A, velocity $= \frac{dx}{dt} = 0$, and $x = a$, from (ii) at A we have,

$$0 = -\mu a^2 + C \Rightarrow C = \mu a^2 \qquad \text{(iii)}$$

Now putting the value of C from (iii) in (ii) we get,

$$\left(\frac{dx}{dt}\right)^2 = -\mu x^2 + \mu a^2 = \mu(a^2 - x^2)$$

$$\Rightarrow \quad v = \frac{dx}{dt} = -\sqrt{\mu}\sqrt{(a^2 - x^2)} \qquad \text{(iv)}$$

Velocity of the particle executing SHM at any instant (–ve sign indicates that x decreases with the increase of time t, i.e. particle moving towards O).

In view of (iv) we have

$$-\frac{dx}{\sqrt{(a^2 - x^2)}} = \sqrt{\mu}\, dt$$

On integrating we get

$$\cos^{-1}\left(\frac{x}{a}\right) = \sqrt{\mu}\, t + C' \qquad \text{(v)}$$

C′ is the constant of integration.

Initially at A t = 0, and $x = a$, from (v) at A we have,

$$\cos^{-1} 1 = \sqrt{\mu} \times 0 + C' \Rightarrow C' = 0$$

Now putting the value of $C' = 0$ in (v) we get,

$$\cos^{-1}\left(\frac{x}{a}\right) = \sqrt{\mu}\, t$$

$$x = a\cos\left(\sqrt{\mu}\, t\right) \qquad \text{(vi)}$$

Position of the particle executing S. H. M at any instant.

§ 2. Characteristics of SHM

1. **At point A**, initial point, $t = 0$ and $x = a$, acceleration $= \frac{d^2x}{dt^2} = -\mu a$, which is maximum numerically, as at all other points $x < a$. Negative sign indicates that it is directed towards O. And the velocity at A is 0, by putting $x = a$, in (iv).
2. **At point O**, $x = 0$, and therefore, from (i) acceleration = 0 and from (iv) the velocity $v = -\sqrt{\mu}\, a$ *which is numerically maximum.* Hence, we observe

that at O though the acceleration is zero but the velocity is not zero thus, the particle will not stop. The velocity is negative therefore the particle will move beyond O.

3. **Beyond O,** when the particle moves beyond O, the velocity of the particle goes on decreasing whereas the acceleration increases and towards O (from (i)). Therefore, the particle will move under retardation and at a stage it will come at rest. Let this point be A′, then from (iv) at A′, we get,

$$0 = -\sqrt{\mu}\sqrt{(a^2 - x^2)} \Rightarrow x = \pm a$$

Consequently, we have

$$OA = a = OA'$$

4. **At point A′,** $x = -a$, acceleration $= \dfrac{d^2x}{dt^2} = \mu a$, which is positive, and is directed towards O i.e. in the direction of x increasing. And the velocity at A' is 0, by putting $x = -a$, in (iv). Thus the particle will begin to move towards O.

Again at O though the acceleration is zero but the velocity is not zero thus, the particle will not stop. The velocity is positive therefore the particle will cross O and move towards A. In this way the particle would keep on oscillating between A and A′, and the particle has two positions **A and A′ where the particle is momentarily at rest.**

Also, from (vi), the time taken from in moving from A to O is given by

$$0 = a\cos\left(\sqrt{\mu}t\right) \Rightarrow \sqrt{\mu}t = \frac{\pi}{2} \text{ or } t = \frac{\pi}{2\sqrt{\mu}}$$

∴ Total time taken by the particle to complete one oscillation, $T = 4 \times$ time taken from in moving from A to O

$$T = 4t = 4\frac{\pi}{2\sqrt{\mu}} = \frac{2\pi}{\sqrt{\mu}} \qquad \text{(vii)}$$

§ 3. Important terminology of SHM

(a) **Amplitude.** The distance from centre O to one of the positions where the particle is momentarily at rest known as amplitude of SHM i.e. AO = A′O = a, is the amplitude of the SHM

(b) **Time period. Total time taken by the particle to complete one oscillation** is known as time period of SHM and it denoted by T, thus we have,

$T = \dfrac{2\pi}{\sqrt{\mu}}$. Time period, which is independent of amplitude.

(c) **Frequency.** The frequency of SHM is the number of oscillations in one second, thus it is the multiplicative inverse of time period.

$$n = \frac{1}{T} = \frac{1}{\frac{2\pi}{\sqrt{\mu}}} = \frac{\sqrt{\mu}}{2\pi}$$

(d) **Epoch.** From (i), we have $\frac{d^2x}{dt^2} + \mu x = 0$, the most general solution of this differential is given by $x = a\cos\left(\sqrt{\mu}t + \alpha\right)$. **The quantity α is known as epoch of SHM**

(e) **Phase.** Phase of the SHM is the time that has elapsed since the particle was at its maximum distance in the positive direction. Now let the particle be at its maximum distance after time t_1, then

$$x = a\cos\left(\sqrt{\mu}t_1 + \alpha\right) = \text{maximum} \Rightarrow \left(\sqrt{\mu}t_1 + \alpha\right) = 0$$

$$\Rightarrow \qquad t_1 = -\alpha/\sqrt{\mu}$$

Hence phase at time $t = t - t_1 = t + \alpha/\sqrt{\mu}$

Important results to be used

1. $\frac{d^2x}{dt^2} = -\mu x$ =Acceleration of the particle executing SHM at any instant and μ is known as intensity of SHM
2. $v = \frac{dx}{dt} = -\sqrt{\mu}\sqrt{(a^2 - x^2)}$ Velocity of the particle executing SHM at any instant
3. $x = a\cos\left(\sqrt{\mu}t\right)$ Position of the particle executing SHM at any instant.
4. Time period $T = \frac{2\pi}{\sqrt{\mu}}$, Frequency $n = \frac{1}{T} = \frac{\sqrt{\mu}}{2\pi}$
5. $v_{max} = \sqrt{\mu}a$ at $x = 0$, Max Acceleration = μa, at $x = a$

ILLUSTRATIVE EXAMPLES

Example 1 If a particle is executing SHM with amplitude *a*. At what distance from the centre will the velocity be half of the maximum.

Solution The velocity of the particle executing SHM at a distance x from the centre is given by

$$v^2 = \mu(a^2 - x^2) \quad \ldots (1)$$

$$v_{max} = \sqrt{\mu}a \quad \ldots (2)$$

As per question $v = v_{max}/2$ and in view of (1) and (2) we have,

$$\mu\left(a^2 - x^2\right) = \frac{1}{4}\mu a^2$$

$$\Rightarrow \qquad 4\left(x^2\right) = 3a^2 \Rightarrow x = \pm\frac{a\sqrt{3}}{2}$$

Example 2 Show that the particle executing SHM requires 1/6 of its time period to move from the position of maximum displacement to one in which the displacement is half the amplitude.

Solution The equation of motion of the particle is given by $\frac{d^2x}{dt^2} = -\mu x$, then

time period of the of the motion is $T = \frac{2\pi}{\sqrt{\mu}}$ $\ldots (1)$

$\frac{dx}{dt} = -\sqrt{\mu}\sqrt{(a^2 - x^2)}$, when the particle moving towards centre O

$$-\frac{dx}{\sqrt{\mu}\sqrt{(a^2 - x^2)}} = dt \quad \ldots (2)$$

let T' time taken by the particle in moving from the position of maximum

displacement i.e. $x = a$

To the position when the displacement is half the amplitude i.e. $x = a/2$. Using (2) we have,

$$T' = \int_a^{a/2} -\frac{dx}{\sqrt{\mu}\sqrt{(a^2 - x^2)}} = \frac{1}{\sqrt{\mu}}\left[\cos^{-1}\left(\frac{x}{a}\right)\right]_a^{a/2} = \frac{1}{\sqrt{\mu}}\left\{\cos^{-1}\frac{1}{2} - \cos^{-1}1\right\}$$

$$\Rightarrow \quad T' = \frac{1}{\sqrt{\mu}}\left\{\frac{\pi}{3} - 0\right\} = \frac{1}{6}T \text{ (using (1)}$$

Example 3 A particle executing SHM of time period T about a centre O and it passes through point P where OP = b with velocity v in the direction of OP, prove that the time in which elapses before it returns to P is $\frac{T}{\pi}\tan^{-1}\left(\frac{vT}{2\pi b}\right)$.

Solution The equation of motion of the particle is given by $\frac{d^2x}{dt^2} = -\mu x$, then time period of the of the motion is

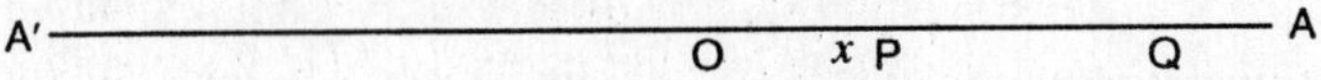

Fig. 5.2

$$T = \frac{2\pi}{\sqrt{\mu}} \Rightarrow \sqrt{\mu} = \frac{2\pi}{T} \qquad \ldots (1)$$

$\frac{dx}{dt} = \sqrt{\mu}\sqrt{(a^2 - x^2)}$, when the particle moving from centre O to A

$$\text{Velocity at P} = v = \sqrt{\mu}\sqrt{(a^2 - b^2)} = \frac{2\pi}{T}\sqrt{(a^2 - b^2)}$$

$$\Rightarrow \quad \sqrt{(a^2 - b^2)} = \frac{vT}{2\pi} \qquad \ldots (2)$$

$$\frac{dx}{\sqrt{\mu}\sqrt{(a^2 - x^2)}} = dt \qquad \ldots (3)$$

let T' time taken by the particle in moving from the position P (OP = b) to maximum displacement i.e. $x = a$

Using (3) we have,

$$T' = \int_b^a \frac{dx}{\sqrt{\mu}\sqrt{(a^2 - x^2)}} = \frac{1}{\sqrt{\mu}}\left[\sin^{-1}\left(\frac{x}{a}\right)\right]_b^a = \frac{1}{\sqrt{\mu}}\left\{\sin^{-1} 1 - \sin^{-1}(b/a)\right\}$$

$$\Rightarrow \quad T' = \frac{1}{\sqrt{\mu}}\cos^{-1}(b/a)$$

$$As\left\{\sin^{-1} 1 - \sin^{-1}(b/a)\right\} = \cos^{-1}(b/a)$$

$$T' = \frac{1}{\sqrt{\mu}}\tan^{-1}\left(\frac{\sqrt{a^2 - b^2}}{b}\right) \qquad \ldots (4)$$

$$\text{If } \cos\theta = \frac{b}{a} \Rightarrow \sin\theta = \frac{\sqrt{a^2 - b^2}}{a} = \tan\theta = \frac{\sqrt{a^2 - b^2}}{b}$$

$$\theta = \cos^{-1}\frac{b}{a} = \tan^{-1}\frac{\sqrt{a^2 - b^2}}{b}$$

Now using (1), (2) and (4) we get,

$$T' = \frac{T}{2\pi}\tan^{-1}\left(\frac{vT}{2\pi b}\right)$$

$$\text{Required time} = 2T' = \frac{T}{\pi}\tan^{-1}\left(\frac{vT}{2\pi b}\right)$$

Example 4 A particle moves in a straight line and its velocity at a distance x from the origin is $k\sqrt{(a^2 - x^2)}$, where k and a are constants. Prove that the motion is SHM and find the amplitude and the time period of the motion.

Solution Given $v^2 = k^2(a^2 - x^2)$ (1)

Differentiating (1) w.r. t. x, we get

$$2v\frac{dv}{dx} = -2k^2x \Rightarrow v\frac{dv}{dx} = -k^2x \qquad \dots (2)$$

As $$v\frac{dv}{dx} = \frac{d^2x}{dt^2} \Rightarrow \frac{d^2x}{dt^2} = -k^2x \qquad \dots (3)$$

Equation (3) is the standard form of SHM, where $\mu = k^2$

Time period $T = \dfrac{2\pi}{\sqrt{k^2}} = \dfrac{2\pi}{k}$

The amplitude point velocity to be zero, thus in view of (1) we have $0 = k^2(a^2 - x^2) \Rightarrow x = a$. Thus amplitude is a.

Example 5 A particle executing SHM of time period T and amplitude a, the velocity v at a distance x from the centre is given by the relation $v^2T^2 = 4\pi^2(a^2 - x^2)$.

Solution The equation of motion of the particle is given by, where O is the centre $\dfrac{d^2x}{dt^2} = -\mu x$, then time period of the motion is

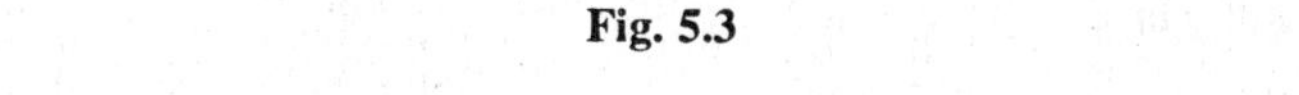

Fig. 5.3

$$T = \frac{2\pi}{\sqrt{\mu}} \Rightarrow \sqrt{\mu} = \frac{2\pi}{T} \qquad \dots (1)$$

$\frac{dx}{dt} = \sqrt{\mu}\sqrt{(a^2 - x^2)}$, when the particle moving from centre O to A

and v be velocity at P

$$\Rightarrow \quad \sqrt{(a^2 - x^2)} = \frac{vT}{2\pi} \qquad \dots (2)$$

On squaring (2) we get,

$$(a^2 - x^2) = \frac{v^2T^2}{4\pi^2} \Rightarrow v^2T^2 = 4\pi^2\left(a^2 - x^2\right)$$

Example 6 A particle executing SHM, if its distances from the middle point of its path at three consecutive seconds are x_1, x_2, x_3. Prove that the time period of SHM is $\frac{2\pi}{\cos^{-1}\left((x_1 + x_3)/2x_2\right)}$.

Solution The equation of motion of the particle is given by, where O is the middle point $\frac{d^2x}{dt^2} = -\mu x$

Let the paticle be at distances x_1, x_2 and x_3 from the middle point O of its path after three consecutive seconds say, t, $(t+1)$ and $(t+2)$ seconds respectively. Then by virtue of position equation we have,

$$x_1 = a\cos\left(\sqrt{\mu}t\right) \qquad \dots (1)$$

$$x_2 = a\cos\left[\sqrt{\mu}(t+1)\right] \qquad \dots (2)$$

$$x_3 = a\cos\left[\sqrt{\mu}(t+2)\right] \qquad \dots (3)$$

Adding (1) and (3) we get

$$x_1 + x_3 = a\cos\left(\sqrt{\mu}t\right) + a\cos\left[\sqrt{\mu}(t+2)\right]$$

$$= 2a\cos\left[\sqrt{\mu}(t+1)\right]\cos\left(\sqrt{\mu}\right)$$

$$\Rightarrow \quad \frac{x_1 + x_3}{2x_2} = \cos\left(\sqrt{\mu}\right) \text{ using (2)}$$

$$\Rightarrow \quad \sqrt{\mu} = \cos^{-1}\left((x_1 + x_3)/2x_2\right) \qquad \dots (4)$$

$$\text{Time period } T = \frac{2\pi}{\sqrt{\mu}} = \frac{2\pi}{\cos^{-1}\left((x_1 + x_3)/2x_2\right)}$$

Example 7 A particle executing SHM, such that at two of its positions the velocities are u and v and corresponding accelerations are α and β; show that the distance between the positions is $\frac{(v^2-u^2)}{(\alpha+\beta)}$. Also find the amplitude of the motion.

Solution The equation of motion of the particle is given by, where O is the middle point

$$\frac{d^2x}{dt^2}=-\mu x \quad \ldots (1)$$

$$V^2=\mu(a^2-x^2) \quad \ldots (2)$$

Here a is amplitude of SHM

Let the particle be at distances x_1 and x_2 from the middle point O of its path having velocities are u and v and corresponding accelerations are α and β respectively. Thus in view of (1) and (2) we have

$$\alpha=-\mu x_1 \quad \ldots (3)$$

$$\beta=-\mu x_2 \quad \ldots (4)$$

$$u^2=\mu\left(a^2-x_1^2\right) \quad \ldots (5)$$

$$v^2=\mu\left(a^2-x_2^2\right) \quad \ldots (6)$$

Adding (3) and (4) and subtracting (5) from (6) we get,

$$(\alpha+\beta)=-\mu(x_1+x_2) \quad \ldots (7)$$

$$(v^2-u^2)=\mu(x_1{}^2-x_2{}^2) \quad \ldots (8)$$

Dividing (8) by (7) we get

$$\frac{(v^2-u^2)}{(\alpha+\beta)}=x_2-x_1 = \text{The required distance between two points}$$

$$\frac{(v^2-u^2)}{(\beta^2-\alpha^2)}=-\frac{1}{\mu} \quad \ldots (9)$$

From (5) and (3) we have

$$a^2=\frac{u^2}{\mu}+x_1{}^2=\frac{u^2}{\mu}+\frac{\alpha^2}{\mu^2}=\frac{1}{\mu^2}\left[u^2\mu+\alpha^2\right] \quad \ldots (10)$$

Now using (9) and (10) we get

$$a = \frac{\sqrt{\{(v^2 - u^2)(v^2\alpha^2 - u^2\beta^2)\}}}{(\beta^2 - \alpha^2)}$$

Example 8 A particle executing SHM, such that the amplitude is *a* and T is time period of the motion. Find the expression of the velocity V in terms of *a* T and *x* also prove that $\int_0^T V^2 dt = \frac{2\pi^2 a^2}{T}$.

Solution The equation of SHM of the particle is given by, where O is the middle point

$$\frac{d^2x}{dt^2} = -\mu x \quad \ldots (1)$$

$$V^2 = \mu(a^2 - x^2) \quad \ldots (2)$$

Here a is amplitude of SHM.

$$T = \frac{2\pi}{\sqrt{\mu}} \Rightarrow \mu = \frac{4\pi^2}{T^2} \quad \ldots (3)$$

$$x = a\cos(\sqrt{\mu}t) \quad \ldots (4)$$

In view of (2) and (3) we get

$$V^2 = \mu(a^2 - x^2) = \frac{4\pi^2}{T^2}(a^2 - x^2) \quad \ldots (5)$$

In view of (3) and (4), (5) reduces to

$$V^2 = \frac{4\pi^2}{T^2}(a^2 - x^2) = \frac{4\pi^2}{T^2}\left(a^2 - \left(a\cos(\sqrt{\mu}t)\right)^2\right)$$

$$V^2 = \frac{4\pi^2}{T^2}(a^2 - x^2) = \frac{4\pi^2}{T^2}\left(a^2 \sin^2 \frac{2\pi}{T}t\right)$$

$$\int_0^T V^2 dt = \int_0^T \frac{4\pi^2}{T^2}\left(a^2 \sin^2 \frac{2\pi}{T}t\right)dt = \frac{2\pi^2 a^2}{T^2}$$

$$\int_0^T \left(1 - \cos\frac{4\pi}{T}t\right)dt = \frac{2\pi^2 a^2}{T^2}\left[t - \frac{T}{4\pi}\sin\frac{4\pi}{T}t\right]_0^T$$

$$\int_0^T V^2 dt = \frac{2\pi^2 a^2}{T} \quad \text{As } \sin 4\pi = \sin 0 = 0$$

Example 9 A particle of mass *m* is attached to a light wire which is stretched tightly between two fixed points with a tension T. If *a* and *b* be the distances of the particle from two ends, prove that the period of small transverse oscillation of mass *m* is $2\pi\sqrt{\dfrac{mab}{T(a+b)}}$.

Solution Let A and B be two given points. The light wire is attached between A and B, C is the point where a particle of mass m is attached. AC = a and BC = b.

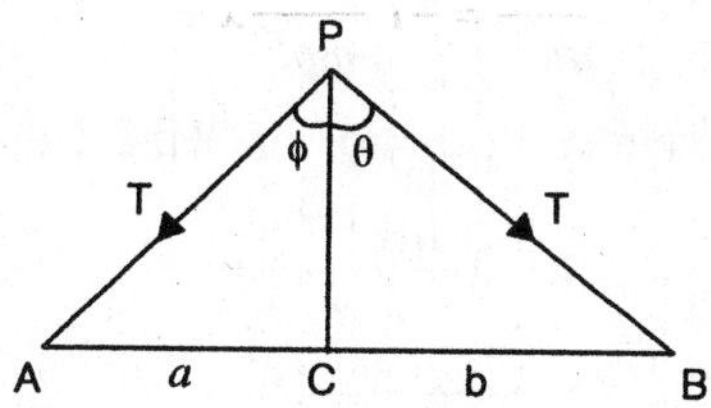

Fig. 5.4

Let the particle be allowed to small transverse displacement **(a displacement which is at right angle to AB)**. Since the displacement is very small therefore, the tension on string remains same. Let P be the position of the particle at time *t*, so that CP = x

The resultant force on the particle along the direction of motion i.e. along PC $= F = T\cos\phi + T\cos\theta$

where $\angle BPC = \theta$ and $\angle APC = \phi$ and $\cos\phi = \dfrac{x}{\sqrt{a^2+x^2}}$

and $\cos\theta = \dfrac{x}{\sqrt{b^2+x^2}}$.

$$F = T\frac{x}{\sqrt{a^2+x^2}} + T\frac{x}{\sqrt{b^2+x^2}}$$

$$\Rightarrow \quad F = Tx\left\{\frac{1}{a}\left(1+\frac{x^2}{a^2}\right)^{-1/2} + \frac{1}{b}\left(1+\frac{x^2}{b^2}\right)^{-1/2}\right\}$$

$$\Rightarrow \quad F = Tx\left\{\frac{1}{a}\left(1 - \frac{x^2}{2a^2} + \ldots\right) + \frac{1}{b}\left(1 - \frac{x^2}{2b^2} + \ldots\right)\right\}$$

$F = Tx\left\{\frac{1}{a} + \frac{1}{b}\right\}$ Since x is very small neglecting the higher powers of x.

$$F = Tx\frac{a+b}{ab} \quad \ldots (1)$$

The equation of motion is given by $m\frac{d^2x}{dt^2} = -Tx\frac{a+b}{ab}$

$$\Rightarrow \quad \frac{d^2x}{dt^2} = -T\frac{a+b}{mab}x \quad \ldots (2)$$

Equation (2) is the standard equation of SHM. Hence

$$\mu = T\frac{a+b}{mab} \quad \ldots (3)$$

$$\text{Time period} = \frac{2\pi}{\sqrt{\mu}} = 2\pi\sqrt{\frac{mab}{T(a+b)}}$$

Example 9 A particle starts from rest under the acceleration k^2x directed towards affixed point and after time t another particle starts from the same position under the same acceleration. Show that the particle will collide at time $\left(\frac{\pi}{k} + \frac{t}{2}\right)$ after the start of the first particle provided $t < \frac{2\pi}{k}$.

Solution The equation of motion of the particle is given by $\frac{d^2x}{dt^2} = -k^2x$, then time period of the of the motion is

$$T = \frac{2\pi}{\sqrt{\mu}} = \frac{2\pi}{k} \quad \ldots (1)$$

$$t < \frac{2\pi}{k}$$

i.e. $t <$ time period, therefore the second particle starts before the first particle has completed first oscillation. Let two particles collide after time t' of the start of second particle.

Therefore, before collision the first particle was in motion for a time $t + t'$
Collision will take place if

Distance travelled by first particle in time $t + t'$ = Distance travelled by 2nd particle in time t'

$$x = a\cos k(t+t') = a\cos kt' \quad \dots (2)$$

$$\Rightarrow \quad k(t+t') = kt' \quad \text{not possible}$$

$$\Rightarrow \quad k(t+t') = 2\pi - kt'$$

$$\Rightarrow \quad t' = \frac{\pi}{k} - \frac{t}{2} \quad \dots (3)$$

$$\text{Required time} = t + t' = t + \frac{\pi}{k} - \frac{t}{2} = \left(\frac{\pi}{k} + \frac{t}{2}\right)$$

REVIEW EXERCISE 5.1

1. The speed v of a particle which moves in a straight line is given by the relation $v^2 = a + 2bx - cx^2$, where x is the distance of the particle at instant t from a fixed point on the path, a, b and c are costants. Show that the motion of the particle is SHM Also find the time period.

   (**Hint.** Acceleration $= v\dfrac{dv}{dx}$, compare with SHM equation, $\mu = c$.

2. Show that if the displacement of a particle moving in a straight line is expressed by the equation $x = a\cos kt + b\sin kt$, it describes a SHM whose amplitude is $\sqrt{a^2 + b^2}$ and period is $2\pi/k$.

   (**Hint.** Find $\dfrac{d^2x}{dt^2} = -k^2x$, for finding amplitude put $\dfrac{dx}{dt} = 0$, we get $\tan kt = b/a$ and substitute in x.

3. A point is moving in a straight line with SHM about a fixed point O on the line, the point has a velocities v_1 and v_2 when the displacements are x_1 and x_2 respectively, form O. Show that the period of motion is

   $$T = 2\pi\sqrt{\frac{x_1^2 - x_2^2}{v_2^2 - v_1^2}}.$$

   (**Hint.** Find $v_2^2 - v_1^2 = \mu$, $x_1^2 - x_2^2$) (and $T = \dfrac{2\pi}{\sqrt{\mu}}$).

4. A particle of mass *m* is attracted towards a fixed point O with a force *mk* times the distance from O. If initially it is projected with velocity *v* towards O from a point distant *c* from O, find the amplitude of oscillation.

   (**Hint.** Eqn of motion is $m\frac{d^2x}{dt^2} = -mkx$, find $\frac{dx}{dt}$ and apply initial condition to obtain constant of integration, for finding amplitude put $\frac{dx}{dt} = 0$.

5. If in a *u, v, w* SHM be the velocities at a distances *a, b, c* respectively from affixed point on the straight line which is not the centre of force; show that the time period of SHM is given by

$$\left(\frac{4\pi^2}{T^2}\right)(b-c)(c-a)(a-b) = \begin{vmatrix} u^2 & v^2 & w^2 \\ a & b & c \\ 1 & 1 & 1 \end{vmatrix}.$$

6 Elastic String (Horizontal and Vertical)

§ 1. Hooke's Law

In an elastic string, the tension is proportional to its extension beyond its natural length. Let l_0 be the natural length and l be the extended length of an elastic string, then the extension produced is $(l - l_0)$ and if T be the tension in the string, then by Hooke's law $T = \frac{\lambda}{l_0}(l - l_0)$, where λ is known as modulus of elasticity.

Theorem. *Prove that the work done against the tension in stretching a light elastic string is equal to the product of its extension and arithmetic mean of initial and final tension.*

Proof. Let l_0 be the natural length. Let it be extended from A to B such that $OA = a$ and $OB = b$.

Fig. 6.1

We have to find the work done against the tension in stretching the string from A to B. Let $OP = x$ and $PQ = \delta x$. Here P and Q are neighboring point i.e. $\delta x \to 0$, the tensions at P and Q are equal. Tension in the string when it stretched upto P is given by

$$T = \frac{\lambda}{l_0}(x - l_0) \quad \text{... (i) acting in the direction of } PO.$$

Work done against the tension in stretching the string from P to Q

$$\delta W = T\delta x = \frac{\lambda}{l_0}(x - l_0)\delta x \qquad \text{(ii)}$$

Work done against the tension in stretching the string from A to B

$$W = \int_a^b \frac{\lambda}{l_0}(x - l_0)\,dx = \frac{\lambda}{l_0}\left[\frac{(x - l_0)^2}{2}\right]_a^b = \frac{\lambda}{2l_0}\left[(b - l_0)^2 - (a - l_0)^2\right]$$

$$W = \frac{\lambda}{2l_0}\left[(b - l_0) - (a - l_0)\right]\left[(b - l_0) + (a - l_0)\right]$$

$$W = [b - a]\frac{1}{2}\left[\frac{\lambda}{l_0}(b - l_0) + \frac{\lambda}{l_0}(a - l_0)\right] \qquad \text{(iii)}$$

From (i) initial tension $T_A = \dfrac{\lambda}{l_0}(a - l_0)$ (iv)

Final tension $T_B = \dfrac{\lambda}{l_0}(b - l_0)$ (v)

In view of (iii), (iv) and (v) we have

$$W = [b - a]\frac{1}{2}[T_A + T_B]$$

= extension produced × mean of initial and final tension

§ 2. Horizontal elastic string. One end of an elastic string whose modulus of elasticity is λ, and natural length is a, is tied to a fixed point on a smooth table, is tied to a fixed point on a smooth horizontal table, and other end is tied to a mass m lying on the table. The particle is pulled to a distance where the extension of the string becomes b, and then let it go; discuss the motion.

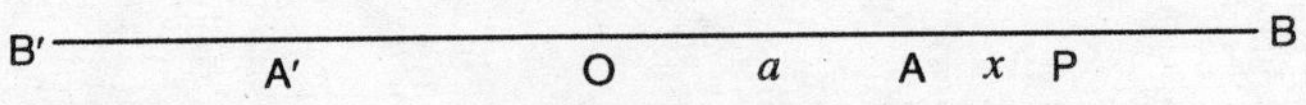

Fig. 6.2

Let OA = *a* be the natural length of the string. Let O be the fixed end and a particle of mass m be attached at A. Let this particle be dragged to B from A such that AB = b and then released. Let P be the position of a particle at any time t and let AP = x.

Tension on the string acting towards O, is given by

$$T = \frac{\lambda}{a}x \qquad \text{(i) (Hooke's law)}$$

The equation of motion of the particle is given by

$$m\frac{d^2x}{dt^2} = -\frac{\lambda}{a}x \Rightarrow \frac{d^2x}{dt^2} = -\frac{\lambda}{ma}x \tag{ii}$$

(ii) is the standard form of SHM.

At B, the velocity of the particle is zero and at A, the acceleration of the particle is zero, thus AB = b is the amplitude of the SHM.

Time period of the motion = $\dfrac{2\pi}{\sqrt{\mu}} = 2\pi\sqrt{\dfrac{ma}{\lambda}}$, (Using (ii))

Time taken by the particle to cover the distance AB,

$$t = \frac{1}{4}\text{ time period} = \frac{1}{4}2\pi\sqrt{\frac{ma}{\lambda}} \Rightarrow t = \frac{1}{2}\pi\sqrt{\frac{ma}{\lambda}} \tag{iii}$$

Multiplying both sides of (ii) by $2\dfrac{dx}{dt}$, we get

$$2\frac{dx}{dt}\frac{d^2x}{dt^2} = -2\frac{dx}{dt}\frac{\lambda}{ma}x \text{ and on integration we get}$$

$$\left(\frac{dx}{dt}\right)^2 = -\frac{\lambda}{ma}x^2 + C$$ (iv) C is the constant of integration.

At point B velocity $= \dfrac{dx}{dt} = 0$ and $x = b$, from (iv) at B we have,

$$0 = -\frac{\lambda}{ma}b^2 + C \Rightarrow C = \frac{\lambda}{ma}b^2 \tag{v}$$

Now putting the value of C from (v) in (iv) we get,

$$\left(\frac{dx}{dt}\right)^2 = -\frac{\lambda}{ma}x^2 + \frac{\lambda}{ma}b^2 = \frac{\lambda}{ma}(b^2 - x^2)$$

$$\Rightarrow \quad v^2 = \frac{\lambda}{ma}(b^2 - x^2) \tag{vi}$$

PointA: When particle reaches at A, then $x = 0$ and the velocity of the particle $v_A = \sqrt{\dfrac{\lambda}{ma}}b$, (putting $x = 0$ in (vi)).

Acceleration at A = 0, (putting $x = 0$ in (ii)).

As soon as the particle reaches at A the string becomes slack i.e. the string gets back to its natural length, hence T = 0 (acceleration is zero), and, therefore, SHM ceases, whereas, the particle has the velocity $\sqrt{\frac{\lambda}{ma}}b$, therefore the particle would move towards O with this velocity till the string gets taut i.e. the particle reaches at point A′ such that OA = OA′ = a.

Motion of the particle between A and A′: Since the string remains slack throughout AA′ i.e. no acceleration is working, *hence the motion of the particle between AA′ is constant velocity.*

Time taken by the particle to cover the distance AO = Time taken by the particle to cover the distance OA′,

$$t' = \frac{\text{Distance}}{\text{velocity}} = \frac{a}{\sqrt{\frac{\lambda}{ma}}b} \Rightarrow t' = \frac{a}{b}\sqrt{\frac{ma}{\lambda}} \qquad \text{(vii)}$$

Time taken by the particle to travel the distance OB

$$= t + t' = \frac{1}{2}\pi\sqrt{\frac{ma}{\lambda}} + \frac{a}{b}\sqrt{\frac{ma}{\lambda}} \quad \text{(using (iii) and (vii))}$$

Required time to complete one oscillation $= 4(t + t')$

$$T = 4\left[\frac{1}{2}\pi\sqrt{\frac{ma}{\lambda}} + \frac{a}{b}\sqrt{\frac{ma}{\lambda}}\right] = 2\sqrt{\frac{ma}{\lambda}}\left(\pi + \frac{2a}{b}\right)$$

ILLUSTRATIVE EXAMPLES

Example 1 A particle of mass m executing SHM in a line joining the points A and B on a smooth table and is contacted with these points by elastic strings whose tensions in equilibrium are each T. Show that the time of an oscillation is $2\pi\sqrt{\frac{mll'}{T(l+l')}}$ · l and l' are the extensions of the strings beyond their natural length.

Solution Let OA and OB be two elastic strings whose natural lengths are a and b respectively. Let O be the position of equilibrium of mass m. Let λ **and λ' be their modulus of elasticity OA and OB respectively.**

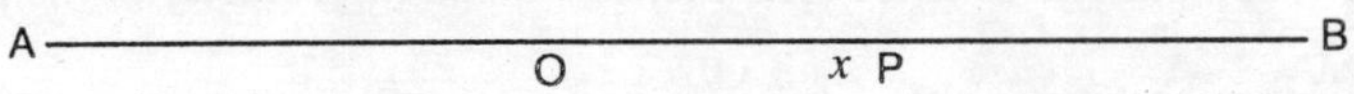

Fig. 6.3

Then in the position of equilibrium, the tension in the string OA = the tension in the string OB i.e.

$$T = \frac{\lambda}{a} l = \frac{\lambda'}{b} l' \qquad \ldots (1)$$

$$\Rightarrow \qquad \frac{\lambda}{a} = \frac{T}{l} \text{ and } \frac{\lambda'}{b} = \frac{T}{l'} \qquad \ldots (2)$$

Let at time t, P be the displaced position of the particle of mass m such that OP = x. Then tension in the string PB $= \frac{\lambda'}{b}(l' - x)$ (acting in the direction of PB and tension in the string PB $= \frac{\lambda'}{b}(l' - x)$ (acting in the direction of PA.

Also the acceleration $\frac{d^2x}{dt^2}$ acting in the direction of OP.

$$m\frac{d^2x}{dt^2} = \frac{\lambda'}{b}(l' - x) - \frac{\lambda}{a}(l + x) = \frac{\lambda'}{b}l' - \frac{\lambda}{a}l - x\left(\frac{\lambda'}{b} + \frac{\lambda}{a}\right) \qquad \ldots (3)$$

In view of (1), (2) and (3) we get,

$$m\frac{d^2x}{dt^2} = -\left(\frac{T}{l'} + \frac{T}{l}\right)x$$

$$\Rightarrow \qquad \frac{d^2x}{dt^2} = -\left(\frac{l + l'}{mll'}\right)Tx, \text{ is the standard equation of SHM}$$

Hence time period $T = \frac{2\pi}{\sqrt{\mu}} = \frac{2\pi}{\sqrt{\left(\frac{l + l'}{mll'}\right)T}} = 2\pi\sqrt{\frac{mll'}{T(l + l')}}$

Example 2 Two elastic strings are fastened to particle of mass m and their other ends to be fixed so that the strings remain taut. λ be their modulus of elasticity, tension T and length *l* and *l′* respectively. Show that the period of oscillation along the line of string is $2\pi\sqrt{\frac{mll'}{(T + \lambda)(l + l')}}$.

Solution Let OA and OB be two elastic strings whose natural lengths are a and b respectively. Let O be the position of equilibrium of mass $m\lambda$. **Their modulus of elasticity of the strings.**

At the fastened state OA = l and OB = l' (given)

Fig. 6.4

Then in the position of equilibrium, the tension in the string OA = the tension in the string OB i.e.

$$T = \frac{\lambda}{a}(l-a) = \frac{\lambda'}{b}(l'-b') \tag{1}$$

$$\Rightarrow \quad \frac{\lambda}{a} l = T + \lambda \text{ and } \frac{\lambda}{b} l' = T + \lambda \tag{2}$$

$$\Rightarrow \quad a = \frac{\lambda l}{T+\lambda} \text{ and } b = \frac{\lambda}{T+\lambda} l' \tag{3}$$

Let at time t, P be the displaced position of the particle of mass m such that OP = x. Then tension in the string PB $= \frac{\lambda}{b}(l'-x-b)$ (acting in the direction of PB and tension in the string PA $= \frac{\lambda}{a}(l+x-a)$ (acting in the direction of PA.

Also the acceleration $\frac{d^2x}{dt^2}$ acting in the direction of OP.

$$m\frac{d^2x}{dt^2} = \frac{\lambda}{b}(l'-x-b) - \frac{\lambda}{a}(l+x-a) = \frac{\lambda}{b}(l'-b) - \frac{\lambda}{a}(l-a) - x\lambda\left(\frac{1}{b}+\frac{1}{a}\right)$$

...(4)

In view of (1), (3) and (4) we get, $m\frac{d^2x}{dt^2} = -\lambda\left(\frac{T+\lambda}{\lambda l'} + \frac{T+\lambda}{\lambda l}\right)x$

$$\Rightarrow \quad \frac{d^2x}{dt^2} = -(T+\lambda)\left(\frac{l+l'}{mll'}\right)x$$, is the standard equation of SHM

Hence time period

$$T = \frac{2\pi}{\sqrt{\mu}} = \frac{2\pi}{\sqrt{(T+\lambda)\left(\frac{l+l'}{mll'}\right)}} = 2\pi\sqrt{\frac{mll'}{(T+\lambda)(l+l')}}$$

Example 3 A light elastic strings whose modulus of elasticity is λ, is stretched to double of its length, and is tied to to two fixed points distance $2a$ apart. A particle of mass m is tied to its middle point in the line of string through a distance equal to half its distance from the fixed point and then released. Find the period of oscillation and the maximum velocity acquired in the subsequent motion.

Solution Let A and B be two fixed points such that AB = $2a$, Let O be its middle point on it such that OA = a, and λ its modulus of elasticity. Let the particle of mass m be displaced from O to C, such that OC = $\frac{a}{2}$. Let at time t, P be the displaced position of the particle of mass m such that OP = x.

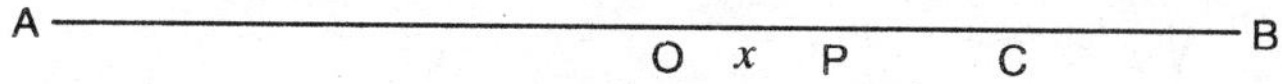

Fig. 6.5

Natural length of OA or OB $= \frac{a}{2}$

The tension in the string AP = T_1

$$T_1 = \frac{\lambda}{a/2}\left(AP - \frac{a}{2}\right) = \frac{\lambda}{a/2}\left((a+x) - \frac{a}{2}\right) = \frac{\lambda}{a/2}\left(\frac{a}{2} + x\right) \qquad \ldots (1)$$

The tension in the string BP = T_2

$$T_2 = \frac{\lambda}{a/2}\left(BP - \frac{a}{2}\right) = \frac{\lambda}{a/2}\left((a-x) - \frac{a}{2}\right) = \frac{\lambda}{a/2}\left(\frac{a}{2} - x\right) \qquad \ldots (2)$$

Also the acceleration $\frac{d^2x}{dt^2}$ acting in the direction of OP.

$$m\frac{d^2x}{dt^2} = T_2 - T_1 = \frac{\lambda}{a/2}\left(\frac{a}{2} - x\right) - \frac{\lambda}{a/2}\left(\frac{a}{2} + x\right) \qquad \ldots (3)$$

$$\Rightarrow \quad m\frac{d^2x}{dt^2} = -\frac{4\lambda}{a}x$$

$\Rightarrow \quad \frac{d^2x}{dt^2} = -\frac{4\lambda}{ma}x$, is the standard equation of SHM

Hence time period $T = \frac{2\pi}{\sqrt{\mu}} = \frac{2\pi}{\sqrt{\frac{4\lambda}{ma}}} = \pi\sqrt{\frac{ma}{\lambda}}$

Amplitude of the SHM = $OC = a/2$.

Maximum velocity of SHM $= \sqrt{\mu} \times \text{Amplitude} = \sqrt{\frac{4\lambda}{ma}} \times \frac{a}{2} = \sqrt{\frac{\lambda a}{m}}$

Example 4 A light elastic strings whose modulus of elasticity is λ and of natural length ($a + b$) and $a > b$. A particle of mass m is attached to it at a distance a from one end say A, which is fixed to a smooth table. Other

end of the string is fixed to a point B so that the string is just unstretched. If the particle is held at B and then released, show that it will oscillate to and fro through a distance $\dfrac{b\left(\sqrt{a}+\sqrt{b}\right)}{\sqrt{a}}$ with time period of $\pi\left(\sqrt{a}+\sqrt{b}\right)\sqrt{\dfrac{m}{\lambda}}$.

Solution Let A and B be two fixed points such that AO = a, and OB = b and let O be point where the mass m be tied and λ, modulus of elasticity of the string. When the particle is held at B, the portion AO of string is stretched whereas the portion OB is slack. When the particle is released from B, the particle wil move towards O due to tension on the portion OA which is stretched upto B. There will be no tension in the portion BO of the string till the particle reaches at O.

Fig. 6.6

Let at time t, P be the displaced position of the particle of mass m such that OP = x, the point P lies between O and B.

Therefore, only tension $T = \dfrac{\lambda}{a}x$ acting toward O at P.

Also the acceleration $\dfrac{d^2x}{dt^2}$ acting in the direction of OP.

$$m\frac{d^2x}{dt^2} = -T = -\frac{\lambda}{a}x \qquad \dots (1)$$

$$\Rightarrow \qquad \frac{d^2x}{dt^2} = -\frac{\lambda}{ma}x, \qquad \dots (2)$$

is the standard equation of SHM

$$\text{Hence time period } T = \frac{2\pi}{\sqrt{\mu}} = \frac{2\pi}{\sqrt{\dfrac{\lambda}{ma}}} = 2\pi\sqrt{\frac{ma}{\lambda}} \qquad \dots (3)$$

$$\text{Time to travel from B to O} = \frac{1}{4}\text{ Time period} = \frac{\pi}{2}\sqrt{\frac{ma}{\lambda}} \qquad \dots (4)$$

Multiplying both sides of (2) by $2\dfrac{dx}{dt}$ we get

$$2\frac{dx}{dt}\frac{d^2x}{dt^2}=-2\frac{dx}{dt}\frac{\lambda}{ma}x$$ and on integration we get

$$\left(\frac{dx}{dt}\right)^2=-\frac{\lambda}{ma}x^2+C \ (5)$$ C is the constant of integration.

At point B velocity $=\frac{dx}{dt}=0$, and $x = b$, from (5) at B we have,

$$0=-\frac{\lambda}{ma}b^2+C\Rightarrow C=\frac{\lambda}{ma}b^2 \qquad \text{... (6)}$$

Now putting the value of C from (6) in (5) we get,

$$\left(\frac{dx}{dt}\right)^2=-\frac{\lambda}{ma}x^2+\frac{\lambda}{ma}b^2=\frac{\lambda}{ma}\left(b^2-x^2\right)$$

$$\Rightarrow \quad b^2-x^2 \qquad \text{... (7)}$$

When particle reaches at O, then $x = 0$

and the velocity of the particle $v_O=\sqrt{\frac{\lambda}{ma}}b$, (putting $x = 0$ in (7).

Acceleration at O = 0, (putting $x = 0$ in (2)).

As soon as the particle crosses O and moves towards left of A, the portion OA becomes slack and OB becomes taut. So the equation of motion becomes

$$\frac{d^2x}{dt^2}=-\frac{\lambda}{mb}x \qquad \text{... (8)}$$

is the standard equation of SHM Hence the particle will come momentarily at rest say the zero velocity point is Q.

$$\text{Hence time period } T=\frac{2\pi}{\sqrt{\mu}}=\frac{2\pi}{\sqrt{\frac{\lambda}{mb}}}=2\pi\sqrt{\frac{ma}{\lambda}} \qquad \text{... (9)}$$

$$\text{Time to travel from O to Q }=\frac{1}{4}\text{ Time period }=\frac{\pi}{2}\sqrt{\frac{mb}{\lambda}} \qquad \text{... (10)}$$

Total time = 2 (time fromB to Q) $=\pi\left(\sqrt{a}+\sqrt{b}\right)\sqrt{\frac{m}{\lambda}}$ (from (4) and (10))

Multiplying both sides of (8) by $2\frac{dx}{dt}$, we get $2\frac{dx}{dt}\frac{d^2x}{dt^2} = -2\frac{dx}{dt}\frac{\lambda}{mb}x$

and on integration we get

$$\left(\frac{dx}{dt}\right)^2 = -\frac{\lambda}{mb}x^2 + C' \qquad \text{... (11)}$$

C′ is the constant of integration.

At point O, velocity $= \frac{dx}{dt} = \sqrt{\frac{\lambda}{ma}}b$ and $x = 0$, from (11) at B we have,

$$\frac{\lambda}{ma}b^2 = -\frac{\lambda}{mb}0 + C' \Rightarrow C' = \frac{\lambda}{ma}b^2 \qquad \text{... (12)}$$

Now putting the value of C′ from (12) in (11) we get,

$$v^2 = \left(\frac{dx}{dt}\right)^2 = -\frac{\lambda}{mb}x^2 + \frac{\lambda}{ma}b^2 \qquad \text{... (13)}$$

At Q, the particle is momentarily at rest, thus $v = \frac{dx}{dt} = 0$. Let at Q $x = d$,

from (13) we have $0 = -\frac{\lambda}{mb}d^2 + \frac{\lambda}{ma}b^2$

$$\Rightarrow \quad d^2 = \frac{b^3}{a} \Rightarrow d = \frac{b\sqrt{b}}{\sqrt{a}} \qquad \text{... (14)}$$

Required distance BQ = BO+OQ $= b + d = b + \frac{b\sqrt{b}}{\sqrt{a}} = \frac{b\left(\sqrt{a}+\sqrt{b}\right)}{\sqrt{a}}$.

REVIEW EXERCISE 6.1

1. A light elastic strings whose modulus of elasticity is λ, is stretched to double of its length, and is tied to two fixed points distance $2l$ apart. A particle of mass m is tied to its middle point in the line of string through a distance equal to half its distance from the fixed point and then released. Find the period of oscillation and the maximum velocity acquired in the subsequent motion. Ans. Max velocity of SHM $= \sqrt{\frac{\lambda l}{m}}$, $T = \pi\sqrt{\frac{ml}{\lambda}}$.
2. A particle of mass m executing SHM in a line joining the points A and B on a smooth table and is contacted with these points by elastic strings

whose tensions in equilibrium are each T; Show that the time of an oscillation is $2\pi\sqrt{\dfrac{mab}{T(a+b)}}$ a and b are the extensions of the strings beyond their natural length.

3. Two elastic strings are fastened to particle of mass m and their other ends to be fixed so that the strings remain taut λ be their modulus of elasticity, tension T and length a and b respectively. Show that the period of oscillation along the line of string is $2\pi\sqrt{\dfrac{mab}{(T+\lambda)(a+b)}}$.

§ 3. Vertical elastic string. *One end of an elastic string whose modulus of elasticity is λ, and natural length is* ***a****, is suspended by one end, to the other end is tied a particle of weight mg, the particle is slightly pulled down and released, discuss the motion.*

Let OA = a be the natural length of the string. Let O be the fixed where the elastic string being suspended. If the particle of weight mg is attached at A to the string, the string get stretched downwards. Let the particle be at B, in the position of equilibrium, such that AB = b.

At B, the particle is in equilibrium, therefore we have,

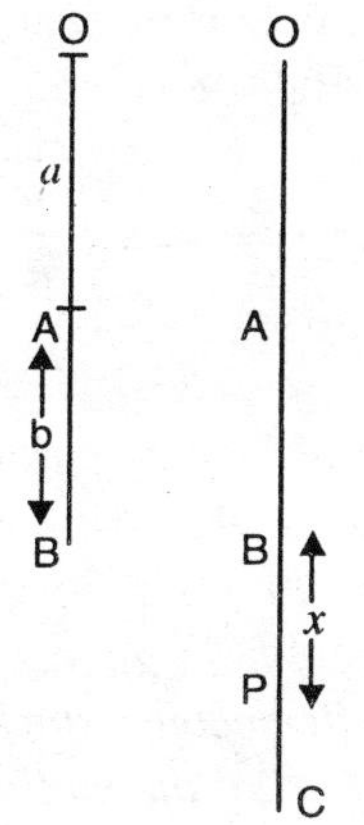

Fig. 6.7

Tension $\dfrac{\lambda}{a}b$ on the string acting vertically upwards = weight of the particle mg acting downwards

Tension on the string acting towards O, is given by

$$T = \frac{\lambda}{a}b = mg \qquad \text{(i)}$$

$$\Rightarrow \qquad \frac{ma}{\lambda} = \frac{b}{g} \qquad \text{(ii)}$$

Now the particle is dragged vertically downwards to C and then released let BC = c.

At C the tension in the string is greater than the weight.

Consequently the particle will begin to move vertically upwards.

Let P be the position of a particle at any time t and let BP = x.

Tension on the string acting vertically upwards O, is given by

$$T = \frac{\lambda}{a}(b + x) \qquad \text{(iii) (Hooke's law)}$$

The equation of motion of the particle is given by

$$m\frac{d^2x}{dt^2} = mg - \frac{\lambda}{a}(b + x) = mg - \frac{\lambda}{a}b - \frac{\lambda}{a}x$$

Using (i) we get,

$$\Rightarrow \qquad \frac{d^2x}{dt^2} = -\frac{\lambda}{ma}x \qquad \text{(iv)}$$

(iv) is the standard form of SHM P

At C, the velocity of the particle is zero and at B, the acceleration C of the particle is zero, thus BC = c is the amplitude of the SHM

$$\text{Time period of the motion} = \frac{2\pi}{\sqrt{\mu}} = 2\pi\sqrt{\frac{ma}{\lambda}} = 2\pi\sqrt{\frac{b}{g}} \quad \text{(v), (Using (ii))}$$

Note:

(*1*) *The equation (iv) will govern the equation the motion of the particle if tension in the string exists i.e. the string remain stretched.*

(2) *If BC < AB, the particle will not rise upto A and whole motion is SHM, thereby equation (iv) and (v) hold good.*

(3) *If BC > AB, the string will become slack when it reaches at A on its upwards journey and the part of the motion above A will be free motion under gravity.*

(4) *In case of a spring the law of compression will work as the extension in the string, however, tension will operate even the particle rises above A (Note 2). Hence the motion is SHM throughout the motion.*

ILLUSTRATIVE EXAMPLES

Note: *In all questions of the vertical elastic strings the position of equilibrium must be used.*

Example 1 A particle of mass m hangs from a fixed point by a light spring and given a small vertical displacement. Show that the motion is SHM If l is the length of the spring when the system is in equilibrium and n the frequency of the motion, show that the natural length of the spring is $l - \dfrac{g}{4\pi^2 n^2}$.

Solution Let OA = a be the natural length of the spring. Let O be the fixed where the spring being suspended. The particle of weight mg is attached at A to the spring, the spring get stretched down wards. Let the particle be at B, in the position of equilibrium, such that OB = l, extension produced = $l - a$.

At B, the particle is in equilibrium, therefore we have,

Tension $\dfrac{\lambda}{a}(l-a)$ on the spring acting vertically upwards = weight of the particle mg acting downwards

Tension on the spring acting towards O, is given by

$$T = \frac{\lambda}{a}(l-a) = mg \qquad \dots (1)$$

$$\Rightarrow \quad \frac{ma}{\lambda} = \frac{(l-a)}{g} \qquad \dots (2)$$

Now the particle is dragged vertically downwards to C and then released let BC = c.

At C the tension in the spring is greater than the weight.

Consequently the particle will begin to move vertically upwards.

Let P be the position of a particle at any time t and let BP = x.

Tension on the spring acting vertically upwards O, is given by

$$T = \frac{\lambda}{a}((l-a)+x) \qquad \dots 3 \text{ (Hooke's law)}$$

The equation of motion of the particle is given by

$$m\frac{d^2x}{dt^2} = mg - \frac{\lambda}{a}((l-a)+x) = mg - \frac{\lambda}{a}(l-a) - \frac{\lambda}{a}x$$

Using (1) we get,

$$\Rightarrow \quad \frac{d^2x}{dt^2} = -\frac{\lambda}{ma}x \qquad \dots(4)$$

(4) is the standard form of SHM.

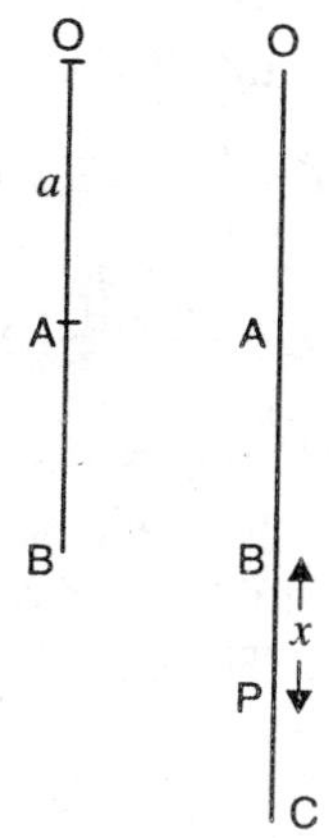

Fig. 6.8

Time period of the motion $t = \frac{2\pi}{\sqrt{\mu}} = 2\pi\sqrt{\frac{ma}{\lambda}} = 2\pi\sqrt{\frac{(l-a)}{g}}$ (5), (Using (2))

Frequency $n = \frac{1}{t} = \frac{1}{2\pi}\sqrt{\frac{g}{(l-a)}}$ ... (6)

On squaring (6) we get, $g = 4\pi^2 n^2 (l-a)$

$\Rightarrow$ Natural length $a = l - \frac{g}{4\pi^2 n^2}$

Example 2 A particle of mass m is attached to one point of a uniform light elastic string. The ends of the string are attached to two points in a vertical lines. Show that the period of vertical oscillation in which the string remains taut is $2\pi\sqrt{\frac{mh}{2\lambda}}$λ, is the coefficient of elasticity and *h* is the harmonic mean of the unstretched length of the two parts.

Solution Let A and B are two points in a vertical line, A to which the two points of the string are attached.

Let O be the point where the particle of mass *m* attached. Let OA = *a* and OB = *b*. Let *l*′ and *l* be the P natural lengths of the portions OA and OB respectively.

There are three forces acting on the particle at O in equilibrium:

(I) Weight *mg* downwards (II) tension $T = \frac{\lambda}{l}(b-l)$ in the string OB vertically downwards (III) tension $T = \frac{\lambda}{l'}(a-l')$ in the string OA vertically upwards.

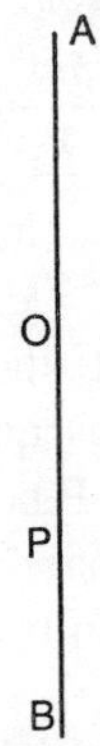

Fig. 6.9

$$\frac{\lambda}{l'}(a-l') = \frac{\lambda}{l}(b-l) + mg$$

(1) As the particle at O in equilibrium.

Now the particle is dragged vertically downwards to B and then released, let OP = *x*. Consequently the particle will begin to move vertically upwards.

The equation of motion of the particle is given by

$$m\frac{d^2x}{dt^2} = mg + \frac{\lambda}{l}((b-x)+l) - \frac{\lambda}{l'}(a+x-l')$$

$$= mg + \frac{\lambda}{l}(b-l) - \frac{\lambda}{l'}(a-l') - \lambda\left(\frac{1}{l}+\frac{1}{l'}\right)x \quad \dots (2)$$

In view of (1) and (2) we get,

$$\Rightarrow \quad \frac{d^2x}{dt^2} = -\frac{\lambda}{m}\left(\frac{1}{l}+\frac{1}{l'}\right)x = -\frac{\lambda}{m}\left(\frac{l+l'}{ll'}\right)x = -\frac{2\lambda}{mh}x \quad \dots (3)$$

Where $h = \dfrac{2ll'}{l+l'}$ = Harmonic mean of l and l' (3) is the standard form of SHM

Time period of the motion $t = \dfrac{2\pi}{\sqrt{\mu}} = 2\pi\sqrt{\dfrac{mh}{2\lambda}}$

Example 3 Two bodies M and M' are attached to a lower end of an elastic string whose upper end is fixed and are hung at rest M', falls off; show that the distance of M from the upper end of the stringat time t is $a+b+c\cos\left\{\sqrt{\left(\frac{g}{b}\right)}\right\}t$, where a is the unstretched length of the string b, and c the distances by which it would be extended when supporting M and M' respectively.

Solution Let OA = a be the natural length of the spring. Let O be the fixed where the spring being suspended. The particle of weight mg is attached at A to the spring, the spring get stretched down wards.

Let the particle be at B, in the position of equilibrium, such that, extension produced = b.

At B, the particle is in equilibrium, therefore we have,

Tension $\dfrac{\lambda}{a}b$ on the spring acting vertically upwards

= weight of the particle mg acting downwards

Tension on the spring acting towards O, is given by

$$\frac{\lambda}{a}b = Mg \quad \dots (1)$$

Similarly for suspension of mass M'

$$\Rightarrow \quad \frac{\lambda}{a}c = M'g \quad \dots (2)$$

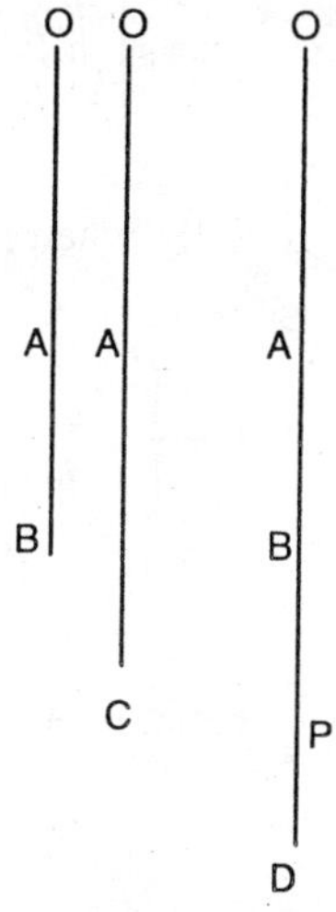

Fig. 6.10

When both the masses are hanging the string will be stretched
So that the extension will be AD = $b + c$.
Now the mass M' falls off, consequently mass M will try to get back
To its equilibrium position of B.
Let P be the position of the mass M at time t such that AP = x
The equation of motion of the particle is given by

$$M\frac{d^2x}{dt^2} = Mg - T = Mg - \frac{\lambda}{a}x$$

Using (1) we get,

$$\Rightarrow \quad \frac{d^2x}{dt^2} = -\frac{g}{b}(x-b) \qquad \dots (3)$$

The solution of (3) is

$$x - b = C_1 \cos\left\{\sqrt{\left(\frac{g}{b}\right)}\right\}t + C_2 \sin\left\{\sqrt{\left(\frac{g}{b}\right)}\right\}t \qquad \dots (4)$$

Initially when $t = 0$, $x = b + c$ putting in (4) we get,

$$b + c - b = C_1 \cos\left\{\sqrt{\left(\frac{g}{b}\right)}\right\}\times 0 + C_2 \sin\left\{\sqrt{\left(\frac{g}{b}\right)}\right\}\times 0$$

$$\Rightarrow \quad C_1 = c \qquad \dots (5)$$

Differentiating (4) we get,

$$\frac{dx}{dt} = \left\{\sqrt{\left(\frac{g}{b}\right)}\right\}\left[-C_1 \sin\left\{\sqrt{\left(\frac{g}{b}\right)}\right\}t + C_2 \cos\left\{\sqrt{\left(\frac{g}{b}\right)}\right\}t\right] \qquad \dots (6)$$

Initially when $t = 0$ velocity $\frac{dx}{dt} = 0$, putting in (6) we get,

$$0 = \left\{\sqrt{\left(\frac{g}{b}\right)}\right\}\left[-C_1 \sin\left\{\sqrt{\left(\frac{g}{b}\right)}\right\}\times 0 + C_2 \cos\left\{\sqrt{\left(\frac{g}{b}\right)}\right\}\times 0\right]$$

$$\Rightarrow \quad C_2 = 0 \qquad \dots (7)$$

In view of (4), (5) and (7) we get $x = b + c\cos\left\{\sqrt{\left(\frac{g}{b}\right)}\right\}t$

Required depth $= a + x = a + b + c\cos\left\{\sqrt{\left(\frac{g}{b}\right)}\right\}t$

Example 4 A light string of natural length l is hung by one end and to the other end is tied successively particles of masses m and m'. If t and t' be the time periods and c and c' statical extensions corresponding to these weights, prove that $g(t^2 - t'^2) = 4\pi^2(c - c')$.

Solution We know that, time period t of vertical elastic string is given by

$t = 2\pi\sqrt{\dfrac{b}{g}}$, where b is the extension.

$t = 2\pi\sqrt{\dfrac{c}{g}}$ and $t' = 2\pi\sqrt{\dfrac{c'}{g}}$ where c and c' are the extensions,

$$\Rightarrow \quad (t^2 - t'^2) = \frac{4\pi^2(c - c')}{g}$$

$$\Rightarrow \quad g(t^2 - t'^2) = 4\pi^2(c - c')$$

Example 5 A light string of natural length a is hung by one end and to the other end is fastened a particle heavy enough to stretch the string to double of its natural length. The string is drawn vertically down till it is four times its natural length and then let it go. Show that the particle return to this point in time $\sqrt{\dfrac{a}{g}}\left(2\sqrt{3} + \dfrac{4\pi}{3}\right)$.

Solution Let OA = a be the natural length of the string. Let O be the fixed where the string being suspended. The particle of weight mg is attached at A to the string, the string get stretched down wards. Let the particle be at B, in the position of equilibrium, such that, OB = $2a$, hence, the extension produced = a

At B, the particle is in equilibrium, therefore we have, Tension $\dfrac{\lambda}{a}a$ on the string acting vertically upwards = weight of the particle mg acting downwards Tension on the spring acting towards O, is given by

$$\frac{\lambda}{a}a = mg \Rightarrow \lambda = mg \qquad \dots (1)$$

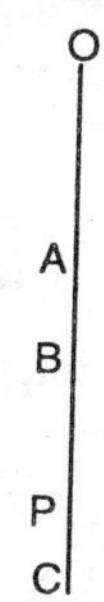

Fig. 6.11

Now the particle is dragged vertically downwards to C such that OC = $4a$ and then released, let OP = x. Consequently the particle will begin to move vertically upwards.

The equation of motion of the particle is given by

$$m\frac{d^2x}{dt^2} = mg - \frac{\lambda}{a}(a + x) \qquad \dots (2)$$

In view of (1) and (2) we have

$$m\frac{d^2x}{dt^2} = mg - \frac{\lambda}{a}(a+x)$$

$$\Rightarrow \quad \frac{d^2x}{dt^2} = -\frac{g}{a}x \qquad \dots (3)$$

Multiplying both sides of (3) by $2\frac{dx}{dt}$, we get

$2\frac{dx}{dt}\frac{d^2x}{dt^2} = -\frac{g}{a}x2\frac{dx}{dt}$ and on integration we get

$$\left(\frac{dx}{dt}\right)^2 = -\frac{g}{a}x^2 + C' \qquad \dots (4)$$

C′ is the constant of integration.

At point C velocity $= \frac{dx}{dt} = 0$, and $x = 2a$, from (4) at C we have,

$$0 = -\frac{g}{a}4a^2 + C' \Rightarrow C' = \frac{g}{a}4a^2 \qquad \dots (5)$$

Now putting the value of C' from (5) in (4) we get,

$$\left(\frac{dx}{dt}\right)^2 = -\frac{g}{a}x^2 + \frac{g}{a}4a^2 = \frac{g}{a}\left(\frac{g}{a}a^2 - x^2\right)$$

$$\Rightarrow \quad v = \frac{dx}{dt} = -\sqrt{\frac{g}{a}}\sqrt{(4a^2 - x^2)} \qquad \dots (6)$$

(–ve sign indicates that x decreases with the increase of time t, i.e. particle moving towards O).

In view of (6) we have

$$-\frac{dx}{\sqrt{(4a^2 - x^2)}} = \sqrt{\frac{g}{a}}dt$$

On integrating we get,

$$\cos^{-1}\left(\frac{x}{2a}\right) = \sqrt{\frac{g}{a}}t + C'' \qquad \dots (7)$$

C″ is the constant of integration.

Initially at C $t = 0$, and $x = 2a$, from (7) at C we have,

$$\cos^{-1} 1 = \sqrt{\frac{g}{a}} \times 0 + C''$$

$$\Rightarrow \quad C'' = 0$$

Now putting the value of C″ = 0 in (7) we get,

$$\Rightarrow \quad t = \sqrt{\frac{a}{g}} \cos^{-1}\left(\frac{x}{2a}\right) \qquad \dots (8)$$

If t' is the time taken in moving from C to A, then putting $x = -a$ in (8) we get

$$t' = \sqrt{\frac{a}{g}} \cos^{-1}\left(\frac{-a}{2a}\right) = \sqrt{\frac{a}{g}} \cos^{-1}\left(\frac{-1}{2}\right)$$

$$\Rightarrow \quad t' = \frac{2\pi}{3}\sqrt{\frac{a}{g}} \qquad \dots (9)$$

Also when the particle reaches A$x = -a$ i.e. we have from (6) the velocity of at A

$$v_A = -\sqrt{\frac{g}{a}}\sqrt{(4a^2 - a^2)} = -\sqrt{3ag}$$

At A string becomes slack and tension vanishes so that the particle moves as free particle under gravity when it rises above A till the string is again taut.

Let the particle rise above A for time t'' till its velocity becomes zero, then from

$$v = u - gt \text{ we get } 0 = \sqrt{3ag} - gt'' \Rightarrow t'' = \sqrt{3\frac{a}{g}} \qquad \dots (10)$$

Time taken by the particle in moving from C to the point where it comes momentarily at rest = $t' + t''$

Required time $T = 2(t' + t'')$

In view of (9) and (10) we get,

$$T = 2\left(\frac{2\pi}{3}\sqrt{\frac{a}{g}} + \sqrt{3\frac{a}{g}}\right) = \sqrt{\frac{a}{g}}\left(2\sqrt{3} + \frac{4\pi}{3}\right)$$

REVIEW EXERCISE 6.2

1. A light string of natural length l is hung by one end and to the other end is tied successively particles of masses m and m'. If T and T' be the time

periods and a and a' statical extensions corresponding to these weights, prove that $g\left(T^2 - T'^2\right) = 4\pi^2(a - a')$.

2. An elastic string without weight of which the unstretched length is l and the modulus of elasticity is the weight of n oz, is suspended by one end and a man of m oz. is attached to the other. Show that the time of a small vertical oscillation is $2\pi\sqrt{\dfrac{ml}{ng}}$.

3. A heavy particle is attached to one end of an elastic string, the other end of which is fixed. The modulus of elasticity of the string is equal to the weight of the particle. The string is drawn vertically down till it is four times its natural length and then let go. Show that the particle will return to this point in time $\sqrt{\dfrac{a}{g}}\left(\dfrac{4\pi}{3} + 2\sqrt{3}\right)$, where is the natural length of the string.

7 Rigid Body Dynamics (Moment of Inertia and Compound Pendulum)

Introduction

In this chapter we shall study the rigid body and its characteristics in terms of moment of inertia and compound pendulum as different shapes and sizes.

§ 1. Rigid Body *A rigid body is a system of particles such that the distance between any two particles remains fixed or invariant.* **Thus any rotation of a rigid body about a fixed point O is equivalent to a rotation about a definite axis through O.**

§ 2. Moment of Inertia (MI). If m be the mass of the particle and r the distance of the particle from a given line, then mr^2 is known as the moment of Inertia of the particle about the given line. It is denoted by I, thus $I = mr^2$

Let a particle of mass m move in a circle of radius r with centre O and angular velocity ω then $v = \omega r$ is the velocity of the particle P. The kinetic energy of the particle

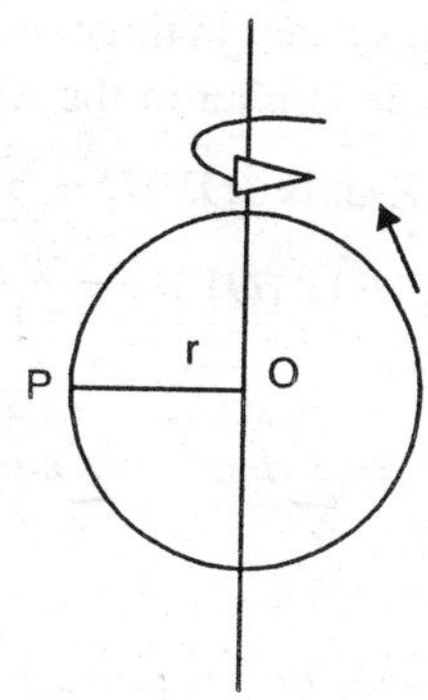

Fig. 7.1

P is given by

$$K.E. = \frac{1}{2}mv^2 = \frac{1}{2}m(\omega r)^2 = \frac{1}{2}mr^2\omega^2 = \frac{1}{2}I\omega^2 \quad ...(1)$$

Where $I = mr^2$ is the moment of inertia of the particle about the given line through O.

Note. *From equation (1) it is clear that the role of moment of inertia in rotational motion is analogous to the mass in linear motion.*

Thus the moment of inertia of a system of particles about a straight line, of masses $m_1, m_2, ..., m_n$ at a distances $r_1, r_2, ..., r_n$ from the same straight line respectively, is defined as

$$I = m_1r_1^2 + m_2r_2^2 + ...m_nr_n^2 = \sum_{i=1}^{i=n} m_ir_i^2$$

Moment of inertia of a continuous distribution of matter in rigid body about a line may be calculated with the help of integration and is defined as

$$I = \int r^2\,dm$$

Where *dm* is the mass of a arbitrary particle of the body and *r* is its distance from the given line.

§ 3. Radius of Gyration. If we express the moment of inertia *I* of a system of particles (body) about a straight line in the form $I = Mk^2$, where *M* is the whole mass of the body, then *k* is known as the radius of gyration of the body about the straight line.

§ 4. Perpendicular Axis Theorem. If I_1 and I_2 are the moments of inertia of a plane lamina about two perpendicular lines on its plane, then the moment of inertia of the lamina about the axis through their intersection and perpendicular to the plane lamina is $I_1 + I_2$.

Let OXYZ be the Cartesian system in three dimensions coordinate system, where O is a point on the plane lamina in the *XY* plane ($z = 0$)

Moment of inertia about X-axis $(OX) I_1 = \sum dmy^2$

Moment of inertia about y-axis $(OY) I_2 = \sum dmx^2$

Moment of inertia about Z-axis

$$(OZ)\, I = \sum dm(x^2 + y^2) = \sum dmx^2 + \sum dmy^2 = I_1 + I_2$$

$$\Rightarrow \quad I = I_1 + I_2.$$

Note *1. Any point on XY plane, the distance of the point from X-axis is y, Y-axis is x and Z-axis is* $\sqrt{x^2 + y^2}$.

2. *The perpendicular axis theorem is applicable to plane lamina only.*

§ 5. Moment of inertia of Some Well known Bodies

(1) Uniform Circular Ring (Hoop)

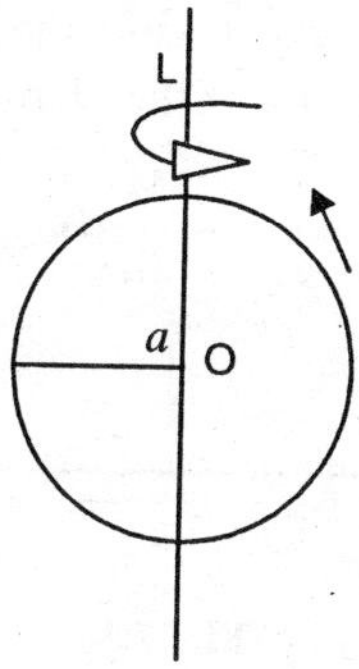

Fig. 7.2

Let the ring be divided into particles of masses $m_1, m_2, \ldots\ m_n$ and their moments of inertia, about the line (OL) through centre O and perpendicular to ring's plane, are

$$m_1a^2,\ m_2a^2, \ldots,\ m_na^2$$

$$MI \text{ of ring about } OL = m_1a^2 + m_2a^2 + \ldots + m_na^2 = (m_1 + m_2 + \ldots m_n)a^2$$

$$MI \text{ of ring about } OL = I = Ma^2 \qquad \ldots (1)$$

As $(m_1 + m_2 + \ldots m_n) = M$.

Let OX and OY be two rectangular axes on the plan of lamina through O.

By symmetry, the MI of the ring about any diameter is the same, and therefore, $I_1 = I_2$ (where I_1 and I_2 are the moments of inertia about OX and OY respectively).

By perpendicular axis theorem, $I = I_1 + I_2 = 2I_1$

$$\Rightarrow \quad I_1 = I_2 = \frac{I}{2} = \frac{1}{2}Ma^2$$

Hence moment of inertia of ring about any diameter $= \frac{1}{2}Ma^2$.

(2) Right circular hallow cylinder. Divide the right circular cylinder into thin circular rings whose planes are perpendicular to line OL. If their masses are $m_1, m_2, \ldots, m_n$ then MI of cylinder about OL

MI of hallow cylinder about OL

$$= m_1a^2 + m_2a^2 + \ldots + m_na^2 = (m_1 + m_2 + \ldots m_n)a^2$$

MI of hallow cylinder about $OL = I = Ma^2$ As $(m_1 + m_2 + ...m_n) = M$.

(3) Rod of length 2*a* and mass *M*. Taking the rod along the X-axis, with the origin at the centre O of the rod, taking the element dx of the rod at the distance x from O and the mass dm of the element dx of the rod then we have,

$$dm = \frac{M}{2a}dx$$

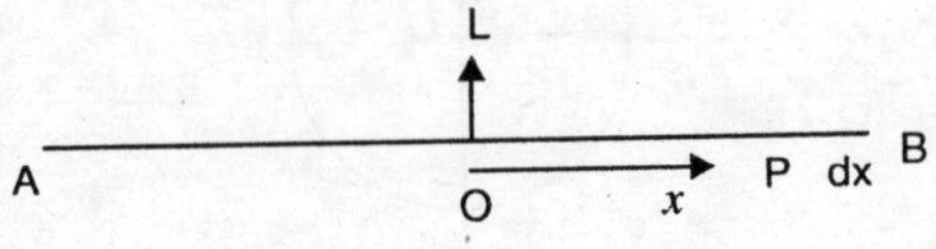

Fig. 7.3

Moment of inertia of the rod about the line through the centre and perpendicular to the rod (OL) is

$$I = \int_{-a}^{a} x^2 \frac{M}{2a} dx = \frac{M}{2a}\left[\frac{x^3}{3}\right]_{-a}^{a} = \frac{M}{2a}\frac{2a^3}{3} = \frac{1}{3}Ma^2$$

Moment of inertia of the rod about the line through one end A and perpendicular to the rod is

$$I = \int_{0}^{2a} x^2 \frac{M}{2a} dx = \frac{M}{2a}\left[\frac{x^3}{3}\right]_{0}^{2a} = \frac{M}{2a}\frac{8a^3}{3} = \frac{4}{3}Ma^2$$

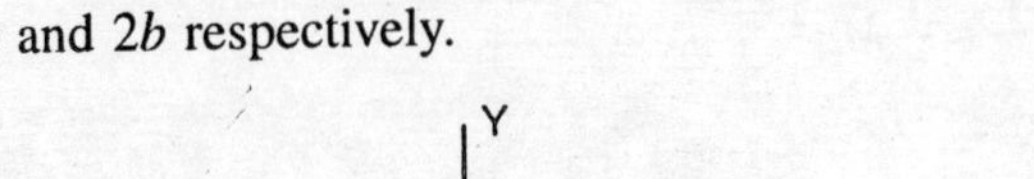

Fig. 7.4

(4) Uniform Rectangular Plate (Lamina) of sides 2*a*, 2*b* and mass *M*. Let O be the centre of mass or CG of the lamina, OX and OY be the lines parallel to the sides $2a$ and $2b$ respectively.

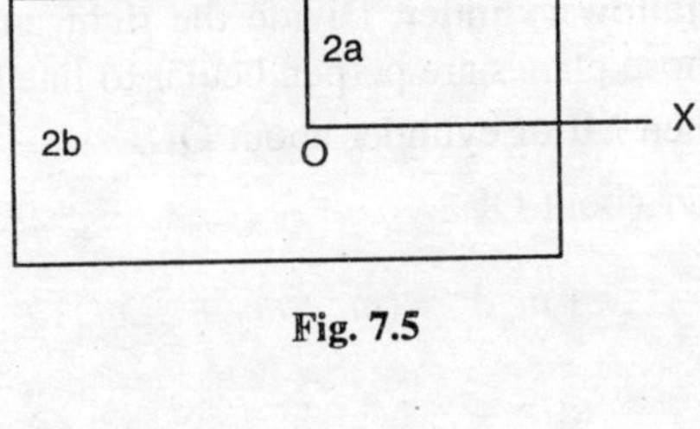

Fig. 7.5

We shall find the MI of the plate about the in its plane passing through the centre O and parallel to the side $2b$ i.e. about the line OY.

The rectangular plate can be considered as the combination of an infinite number of thin strips parallel to the side $2a$. If dm be the mass of one such strip, then the MI of the strip about OY is $\frac{1}{3}a^2dm$. Hence MI of the whole plate about OY is

$$I_1 = \frac{1}{3}Ma^2 \quad \dots (1)$$

Similarly, the moment of inertia (M.I.) of the whole plate about OX is

$$I_2 = \frac{1}{3}Mb^2 \quad \dots (2)$$

By perpendicular axis theorem, the moment of inertia of the rectangular plate about the line through O and perpendicular to its plane is

$$I = I_1 + I_2 = \frac{1}{3}M\left(a^2 + b^2\right) \quad \dots (3)$$

(4) Uniform circular disc Plate (Lamina) of radius *a* and mass *M*. The circular disc plate can be considered as the combination of an infinite number of thin rings with centre O. Consider one such ring whose inner and outer and inner radii are r and $r + dr$, respectively. The area of the ring is $2\pi r dr$ and its mass dm is given by

$$dm = M\frac{2\pi r dr}{\pi a^2} = M\frac{2rdr}{a^2} \quad \dots (1)$$

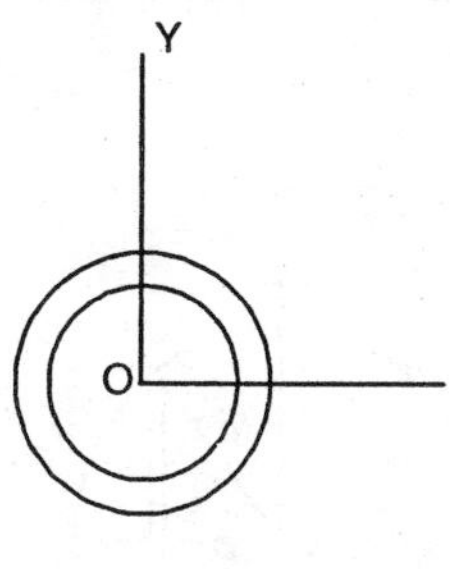

Fig. 7.6

The MI of the ring of mass dm and radius r about any diameter is given by y

$$dI = \frac{1}{2}dmr^2 \quad \dots (2)$$

The MI of the circular disc about any diameter is given by

$$I=\int_0^a \frac{1}{2}dmr^2=\int_0^a \frac{1}{2}M\frac{2rdr}{a^2}r^2=\frac{M}{a^2}\int_0^a r^3dr=\frac{1}{4}Ma^2$$

$\Rightarrow \quad I=\frac{1}{4}Ma^2 \qquad$ (MI about any diameter)

Thus we have,

$$I_1=\frac{1}{4}Ma^2 \qquad \ldots (3)$$

(about the horizontal diameter taken as X-axis)

$$I_2=\frac{1}{4}Ma^2 \qquad \ldots (4)$$

(about the vertical diameter taken as Y-axis)

By perpendicular axis theorem, **the moment of inertia of the circular disc plate about the line through centre O and perpendicular to the circular disc plane is**

$$I=I_1+I_2=\frac{1}{4}M\left(a^2+a^2\right)=\frac{1}{2}Ma^2 \qquad \ldots (5)$$

(5) Uniform solid sphere of radius *a* and mass *M*. The **solid sphere** can be considered as the combination of an infinite number of thin disc by planes perpendicular to the given diameter. Consider one such disc of radius y. Then the mass of the disc of radius y between planes x and $x + dx$ from the centre is

$$dm=M\frac{\pi y^2dx}{\frac{4}{3}\pi a^3}=M\frac{3y^2dx}{4a^3} \quad \ldots(1) \qquad \text{where } y^2=a^2-x^2$$

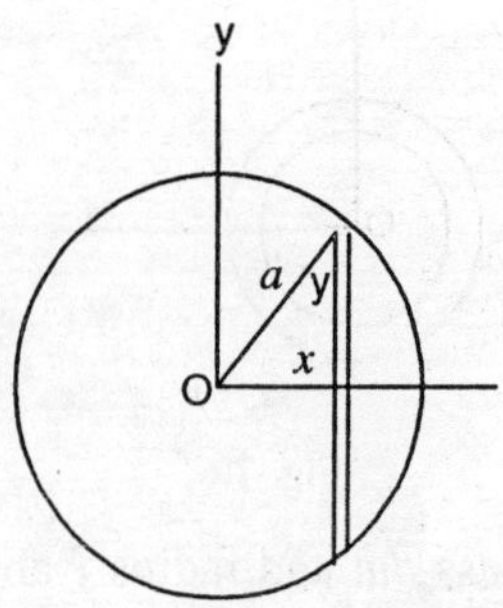

Fig. 7.7

The MI of the disc of mass dm and radius y about the line through the centre and perpendicular to the plane of the disc is

$$dI = \frac{1}{2} dmy^2 = \frac{1}{2} M \frac{3y^4 dx}{4a^3} \qquad \dots (2)$$

The M.I. of the sphere about any diameter is given by

$$I = \int_{-a}^{a} \frac{1}{2} M \frac{3y^4 dx}{4a^3} = \int_{-a}^{a} \frac{1}{2} M \frac{3(a^2 - x^2)^2 dx}{4a^3} = \frac{2}{5} Ma^2 \qquad \dots (3)$$

(5) Uniform spherical shell (hallow sphere) of radius *a* and mass *M*. The **spherical shell** can be considered as the combination of an infinite number of thin ring by planes perpendicular to the given diameter. Consider one such ring of radius y whose centre is at a distance x from the centre of the spherical shell and the breadth of such ring is ds.

Now radius of the ring $y = \sqrt{a^2 - x^2} \Rightarrow \frac{dy}{dx} = \frac{-x}{y}$, ... (1)

Surface area of the ring $= 2\pi y \delta s = 2\pi y \sqrt{1 + \left(\frac{dy}{dx}\right)^2} dx = 2\pi y \sqrt{1 + \left(\frac{-x}{y}\right)^2} dx$

$$= 2\pi y \frac{\sqrt{y^2 + x^2}}{y} dx = 2\pi y \frac{a}{y} dx = 2\pi a dx \qquad \dots (2) \ (\text{as } y^2 + x^2 = a^2)$$

Then the mass of the ring of radius y whose centre is at a distance x from the centre of the spherical shell

$$dm = M \frac{2\pi a dx}{4\pi a^2} = M \frac{dx}{2a} \qquad \dots (3)$$

The MI of the ring of radius y and mass dm about the line through the centre of the ring and perpendicular to the plane of the ring is

$$dI = dmy^2 = M \frac{dx}{2a} y^2 \qquad \dots (4)$$

The MI of the spherical shell about diameter along X axis is given by

$$I = \int_{-a}^{a} M \frac{dx}{2a} y^2 = \int_{-a}^{a} \frac{1}{2a} M (a^2 - x^2) dx = \frac{2}{3} Ma^2 \qquad \dots (5)$$

Another method. Hallow sphere of radius *a* and mass *M*. Hallow sphere is symmetrical about any diameter, therefore MI about any diameter is the same. Taking any three mutually perpendicular diameters as axes of X, Y, Z and If I_1, I_2 and I_3 be the moments of inertia about these axes respectively. We have $I_1 = I_2 = I_3$ (due to symmetry)

$$\therefore \quad 3I_1 = I_1 + I_2 + I_3 = \sum m\left(y^2 + z^2 + z^2 + x^2 + x^2 + y^2\right)$$

$$= 2\sum m\left(x^2 + y^2 + z^2\right)$$

$$\therefore \quad 3I_1 = 2\sum ma^2 = 2Ma^2$$

$$\Rightarrow \quad I = \frac{2}{3}Ma^2 \qquad \text{(MI about any diameter)}$$

§ 4. Parallel Axis Theorem. The moments of inertia of about any axis is equal to its moment of inertia about a parallel axis through its centre of mass (gravity), plus the moment of inertia of the whole mass at the centre of mass (gravity) about the axis.

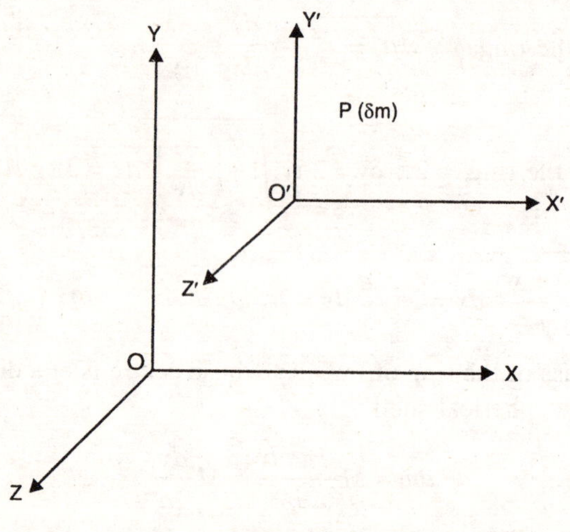

Fig. 7.8

Let OXYZ be rectangular axes at any point O of the body. Let O′ be the centre of mass (gravity) of the rigid body and O′ has the coordinates $(\bar{x}, \bar{y}, \bar{z})$ with respect to O.

Let (x, y, z) and (x', y', z') be the coordinates of an arbitrary particle of mass δm at point P w.r.t. OXYZ and O′X′Y′Z′ respectively.

Thus we have $\quad x = \bar{x} + x', \; y = \bar{y} + y', \; z = \bar{z} + z' \quad \ldots (1)$

The coordinates of the centre of mass O′ w.r.t. OXYZ are

$$\left(\frac{\sum \delta m x'}{\sum \delta m}, \frac{\sum \delta m y'}{\sum \delta m}, \frac{\sum \delta m z'}{\sum \delta m}\right)$$

Since O′ is a new origin, therefore,

$$\frac{\sum \delta m x'}{\sum \delta m} = 0 \Rightarrow \sum \delta m x' = 0 \quad \text{... (2)}$$

$$\frac{\sum \delta m y'}{\sum \delta m} = 0 \Rightarrow \sum \delta m y' = 0 \quad \text{... (3)}$$

$$\frac{\sum \delta m z'}{\sum \delta m} = 0 \Rightarrow \sum \delta m z' = 0 \quad \text{... (4)}$$

The MI of the rigid body about X-axis is

$$I_x = \sum \delta m\left(y^2 + z^2\right) = \sum \delta m\left[\left(\bar{y} + y'\right)^2 + \left(\bar{z} + z'\right)^2\right] \quad \text{(using (1)}$$

$$= \sum \delta m\left[\bar{y}^2 + y'^2 + 2y'\bar{y} + \bar{z}^2 + z'^2 + 2z'\bar{z}\right]$$

$$= \sum \delta m\left(\bar{y}^2 + \bar{z}^2\right) + \sum \delta m\left(y'^2 + z'^2\right) + 2y\sum \delta m\, y' + 2z\sum \delta m z'$$

$$= \left(\bar{y}^2 + \bar{z}^2\right)\sum \delta m + \sum \delta m\left(y'^2 + z'^2\right) \quad \text{(using (3) and (4))}$$

$$= M\left(\bar{y}^2 + \bar{z}^2\right) + \sum \delta m\left(y'^2 + z'^2\right)$$

$$\therefore \quad I_x = M\left(\bar{y}^2 + \bar{z}^2\right) + I_{x'}$$

Where $M = \sum \delta m$ is the total mass of the rigid body $I_{x'} = \sum \delta m\left(y'^2 + z'^2\right)$, is the MI of the body about the axis parallel to X-axis i.e. X' axis passing through the centre of mass O′ and $M\left(\bar{y}^2 + \bar{z}^2\right)$ is the MI of the total mass M concentrated at the centre of mass O′ about X-axis.

Similarly

$$I_y = M\left(\bar{z}^2 + \bar{x}^2\right) + I_{y'} \text{ and } I_z = M\left(\bar{x}^2 + \bar{y}^2\right) + I_{z'}$$

Note. *The general mathematical form of the parallel axes theorem is* $I_O = I_{O'} + Md^2$, *where M is the total mass of the rigid body d, is the distance between the parallel lines through O and the centre of mass (gravity) O′* I_O, *is the MI about a line l and* $I_{O'}$ *is the MI about a line parallel to the line l and passing through the centre of mass O′.*

Important results to be used

1. $I = Mk^2$, where M is the whole mass of the body, then k is radius of gyration of the body.
2. $I = I_1 + I_2$, I_1, I_2 are the MI of a plane lamina about two perp. lines on its plane, and I is MI of the lamina about the axis through their intersection and perp. to the lamina. (Perp. Axis Theorem)
3. $I_O = I_{O'} + Md^2$, where M is the total mass of the rigid body, d is the distance between the parallel lines through O and the centre of mass (gravity) O′ (Parallel Axes Theorem)
4. MI of ring about any diameter $= \frac{1}{2}Ma^2$, MI about the line through centre O and perp. to ring's plane $= Ma^2$.
5. Rod of length $2a$ and mass M, MI about the line through the centre and perp. to the rod $= \frac{1}{3}Ma^2$.
6. M.I of disc about any diameter $= \frac{1}{4}Ma^2$, MI about the line through centre O and perp. to disc plane $= \frac{1}{2}Ma^2$.
7. The MI of the sphere about any diameter $= \frac{2}{5}Ma^2$, The MI of the spherical shell about any diameter $= \frac{2}{3}Ma^2$.

ILLUSTRATIVE EXAMPLES

Applications of Parallel Axis Theorem and Perpendicular Axis Theorem

Example 1 Find the Momemt of Inertia of the following rigid bodies

(i) Rod of length $2a$ and mass M about the line through one end.

(ii) Uniform Rectangular Plate of sides $2a$, $2b$ and mass M about a side of length $2b$ and $2a$.

(iii) Uniform circular disc Plate of radius a and mass M about a tangent line.

Solution (i). MI of the rod about the line through one end A and perpendicular to the rod is

$$I = \int_0^{2a} x^2 \frac{M}{2a} dx = \frac{M}{2a}\left[\frac{x^3}{3}\right]_0^{2a} = \frac{M}{2a}\frac{8a^3}{3} = \frac{4}{3}Ma^2$$

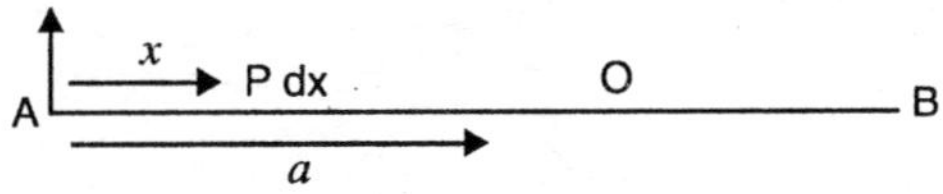

Fig. 7.9

2nd Method. By parallel axis theorem the Moment of inertia of the rod of length $2a$ and mass M about the line through one end is given by

$I_A = I_O + Md^2$ where I_0 is the MI about a line perpendicular to rod and passing through the centre of mass Od, is the distance between O and A.

$$I_A = I_O + Md^2 = \frac{1}{3}Ma^2 + Ma^2 = \frac{4}{3}Ma^2$$

(ii) By parallel axis theorem the Moment of inertia of Uniform Rectangular Plate of sides 2a, $2b$ and mass M about a side of length $2b$ is given by

$$I = I_O + Md^2 = \frac{1}{3}Ma^2 + Ma^2 = \frac{4}{3}Ma^2$$

Similarly, about a side of length $2a$ is given by

$$I = I_O + Md^2 = \frac{1}{3}Mb^2 + Mb^2 = \frac{4}{3}Mb^2$$

(iii) By parallel axis theorem the Moment of inertia of Uniform circular disc Plate of radius a and mass M about a tangent line

$I = I_O + Md^2 = \frac{1}{2}Ma^2 + Ma^2 = \frac{3}{2}Ma^2$, the distance from centre to the tangent line is a.

Example 2 The particles of same mass (m) are fixed at three points on the circumference of a uniform circular hoop (ring) which is of mass M and radius a, so that three masses form an equilateral triangle. The hoop is free to rotated in a vertical plane about the point on the circumference opposite to one of the masses m. Find the MI of the system about the line passing through the point of suspension and perpendicular to the plane.

Solution Let P be the point of suspension and A, B and C be the points on the circumference where the particles are placed, so that ΔABC be an equilateral triangle.

M.I. of the hoop about P is $I_1 = Ma^2 + Ma^2 = 2Ma^2$ ... (1)
APBO is a rhombus, AP = BO = a

M.I. of the mass at C about P is $I_2 = m(2a)^2 = 4ma^2$ ... (2)
BP = AO = a.

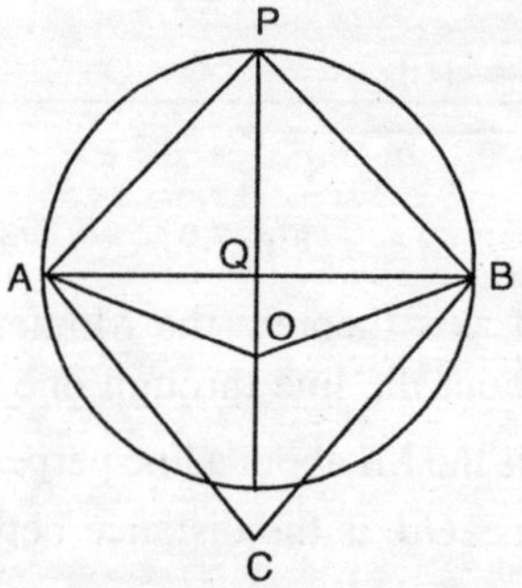

Fig. 7.10

MI of the mass at A about P is $I_3 = m(a)^2 = ma^2$ ... (3)

MI of the mass at B about P is $I_4 = m(a)^2 = ma^2$ ... (4)

Total MI of the system = $I_1 + I_2 + I_3 + I_4 = 2Ma^2 + 6ma^2$

Example 2 A rigid body is ABC consists of two uniform rods AB and BC attached to each other such that ABC is a straight line. AB is of mass m and length 6*a*, BC is of mass 3m and length 2*a*. Find the M.I. of the system about the line passing through the point of suspension A and perpendicular to the plane.

Solution MI of AB about A is

$$I_1 = \frac{1}{3}m(3a)^2 + m(3a)^2 = 12ma^2 \quad ... (1)$$

MI of the mass at BC about A is

$$I_2 = \frac{1}{3}3m(a)^2 + 3m(7a)^2 = 148ma^2 \quad ... (2)$$

Total MI of the system $= I_1 + I_2 = 160ma^2$ ... (3)

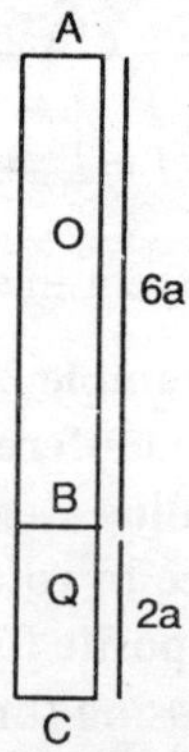

Fig. 7.10

§ 6. Application of Moment of Inertia (Compound Pendulum)

A compound pendulum is any rigid body which oscillates freely about a fixed horizontal axis. No external force acts on a pendulum apart from its weight and supporting force at the axis.

Corollary: *A compound pendulum, of mass m and centre of gravity G, is free to rotated about a fixed horizontal axis through a point A. AG is perpendicular to the axis of rotation and of l⌃..gth h. The moment of inertia of the pendulum* *about a fixed axis (throughA) and parallel through G be I_A and I_G. Find the following:*

(a) Time period of compound pendulum, (b) Length of simple equivalent pendulum (SEP). Length of a simple pendulum of same time period, (c) The time period in terms of radius of gyration.

1st Method (Moment method)

Solution Let *A* be the point of suspension and *G* be the centre of mass of the system of mass *M*, so that $AG = h$.

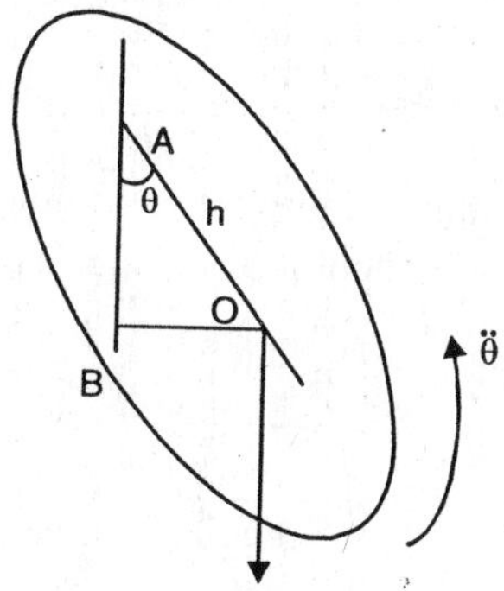

Fig. 7.11

When *AG* makes an angle θ with the down ward direction, now taking the moments about *A*, we have

$$\tau = -Mgh\sin\theta = I_A\ddot{\theta} \qquad \text{(i)}$$

where I_A *is the moment of inertia of the* compound pendulum about A and $\ddot{\theta}$ is the angular acceleration of the body.

Equation (i) can be rewritten as

$$\ddot{\theta} = \frac{-Mgh\sin\theta}{I_A} \qquad \text{(ii)}$$

Hence, for small oscillation i.e. $\theta \to 0$, we get $\ddot{\theta} = -\dfrac{Mgh}{I_A}\theta$ (iii)

Equation (iii) represents Simple Harmonic Motion with time period *T* is given by

$T = 2\pi\sqrt{\dfrac{I_A}{Mgh}}$, by parallel axis theorem $I_A = I_G + Mh^2$

$$T = 2\pi\sqrt{\frac{I_G + Mh^2}{Mgh}}$$ (iv) required equation of time period.

(b) The period for small oscillations of a simple pendulum of length l is given by

$$T = 2\pi\sqrt{\frac{l}{g}} \quad \text{(v)}$$

Comparing (iv) and (v), we have

$$l = \frac{I_G + Mh^2}{Mh} = \frac{I_A}{Mh} \quad \text{(vi)}$$

l, is known as length of simple equivalent pendulum (SEP).

(c) The radius of gyration about axis through G is k, so

$$I_G = Mk^2 \Rightarrow I_A = Mk^2 + Mh^2$$

$$T = 2\pi\sqrt{\frac{Mk^2 + Mh^2}{Mgh}} = 2\pi\sqrt{\frac{k^2 + h^2}{gh}} \quad \text{(vii)}$$

2nd Method (Energy method)

The kinetic energy is given by

$$KE = \frac{1}{2} I_A \dot{\theta}^2 \quad \text{(i)}$$

Considering A as a base level, the potential energy of the system when the centre of mass of the system at G such that $AG = h$ (shown in fig), is given by

$$PE = -Mgh\cos\theta \quad \text{(ii)}$$

By the principle of conservation of energy, the total energy is constant, so that

$$\frac{1}{2} I_A \dot{\theta}^2 + -Mgh\cos\theta = \text{constant} \quad \text{(iii)}$$

Differentiating w.r. t. time t we get,

$$I_A \dot{\theta}\ddot{\theta} + Mgh\sin\theta\dot{\theta} = 0 \quad \text{As} \frac{d}{dt}\dot{\theta}^2 = 2\dot{\theta}\ddot{\theta}$$

Dividing by $\dot{\theta}$ and rearranging, we get,

$$\ddot{\theta} = \frac{-Mgh\sin\theta}{I_A} \quad \text{(iv)}$$

Hence, for small oscillation i.e. $\theta \to 0$ we get $\ddot{\theta} = -\frac{Mgh}{I_A}\theta$, (v)

Equation (vi) represents Simple Harmonic Motion with time period T is given by

$$T = 2\pi\sqrt{\frac{I_A}{Mgh}}, \text{ by parallel axis theorem } I_A = I_G + Mh^2$$

$$T = 2\pi\sqrt{\frac{I_G + Mh^2}{Mgh}}$$ (vii) required equation of time period.

Important results to be used

1. MI of ring about any diameter $= \frac{1}{2}Ma^2$, MI about the line through centre O and perp. to ring's plane $= Ma^2$.
2. Rod of length $2a$ and mass M, MI about the line through the centre and perp. to the rod $= \frac{1}{3}Ma^2$.
3. MI of disc about any diameter $= \frac{1}{4}Ma^2$, MI about the line through centre O and perp. to disc plane $= \frac{1}{2}Ma^2$.
4. The MI of the sphere about any diameter $= \frac{2}{5}Ma^2$, The MI of the spherical shell about any diameter $= \frac{2}{3}Ma^2$
5. Time period $T = 2\pi\sqrt{\frac{I_A}{Mgh}} = 2\pi\sqrt{\frac{I_G + Mh^2}{Mgh}} = 2\pi\sqrt{\frac{k^2 + h^2}{gh}}$
6. length of simple equivalent pendulum (SEP)
$l = \frac{I_G + Mh^2}{Mh} = \frac{I_A}{Mh} = \frac{k^2 + h^2}{h}$, k is radius of gyration and h is the distance between point of suspension A and C.G of the system.

ILLUSTRATIVE EXAMPLES

Applications of Moment of Inertia in Compound Pendulum

Example 1 The particles of same mass (m) are fixed at three points on the circumference of a uniform circular hoop (ring) which is of mass M and radius a, so that three masses form an equilateral triangle. The hoop is free to rotate in a vertical plane about the point on the circumference opposite to one of the masses m. Find the SEP length of the system and time period.

Solution Let P be the point of suspension and A, B and C be the points on the circumference where the particles are placed, so that ΔABC be an equilateral triangle.

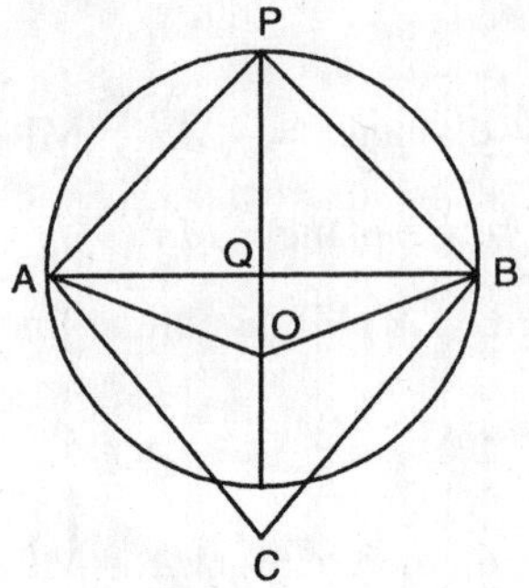

Fig. 7.12

MI of the hoop about P is $I_1 = Ma^2 + Ma^2 = 2Ma^2$

$APBO$ is a rhombus, $AP = BO = a$

MI of the mass at C about P is $I_2 = m(2a)^2 = 4ma^2$

$BP = AO = a$

MI of the mass at A about P is $I_3 = m(a)^2 = ma^2$

MI of the mass at B about P is $I_4 = m(a)^2 = ma^2$

Total MI of the system $= I_1 + I_2 + I_3 + I_4 = 2Ma^2 + 6ma^2$ ... (1)

Since circumcentre of an equilateral triangle is same as that of C.

Centre of gravity of the equilateral triangle, hence CG of the system lies at the centre of the hoop. Therefore, we have, $h = OP = a$

$$\text{SEP length } l = \frac{\text{Total M.I. of the system}}{\text{OP} \times \text{total mass of the system}} = \frac{2Ma^2 + 6ma^2}{a \times (M + 3m)} = 2a$$

$$\text{Time period } T = 2\pi\sqrt{\frac{l}{g}} = 2\pi\sqrt{\frac{2a}{g}}$$

Example 2 A rigid body is ABC consists of two uniform rods AB and BC attached to each other such that ABC is a straight line. AB is of mass m and length 6*a*, BC is of mass 3m and length 2*a*. If the system is free to rotate in a vertical plane about the point A. Find the SEP length of the compound pendulum and time period.

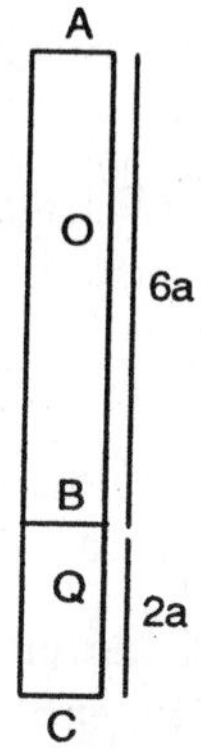

Fig. 7.13

Solution MI of AB about A is $I_1 = \frac{1}{3}m(3a)^2 + m(3a)^2 = 12ma^2$

MI of BC about A is $I_2 = \frac{1}{3}3m(a)^2 + 3m(7a)^2 = 148ma^2$

Total MI of the system $= I_1 + I_2 = 160ma^2$ ...(1)

The distance between CG and the point of suspension A is

$$h = \frac{\text{Mass of AB} \times AO + \text{Mass of BC} \times AQ}{\text{Total mass}}$$

$$\Rightarrow \quad h = \frac{m \times 3a + 3m \times 7a}{3m + m} = 6a = AB \qquad ...(2)$$

In view of (1) and (2) we get,

$$\text{SEP length } l = \frac{\text{Total M.I. of the system}}{h \times \text{total mass of the system}} = \frac{160ma^2}{6a \times 4m} = 20a/3$$

$$\text{Time period } T = 2\pi\sqrt{\frac{l}{g}} = 2\pi\sqrt{\frac{20a}{3g}}$$

Example 3 A compound pendulum consists of a uniform rod AB of mass M and length 2*l*, at the end point of the rod a uniform circular disc of mass 3M and of radius 2*l*/3 is attached in such a way that centre of the disc and B coincide. If the system is free to rotate about horizontal axis which passes through the pointA, find the SEP length of the compound pendulum and time period.

Solution MI of AB about A is $I_1 = \frac{1}{3}M(l)^2 + M(l)^2 = \frac{4}{3}Ml^2$

MI of the disc about A is $I_2 = \frac{1}{2}3M\left(\frac{2l}{3}\right)^2 + 3M(2l)^2 = \frac{38}{3}Ml^2$

Total MI of the system $= I_1 + I_2 = 14Ml^2$ ...(1)

The distance between CG and the point of suspension A is

$$h = \frac{\text{Mass of AB} \times \text{AO} + \text{Mass of disc} \times \text{AB}}{\text{Total mass}}$$

$$\Rightarrow \quad h = \frac{M \times l + 3M \times 2l}{M + 3M} = \frac{7}{4}l \qquad \ldots (2)$$

In view of (1) and (2) we get,

$$\text{SEP length } L = \frac{\text{Total M.I. of the system}}{h \times \text{total mass of the system}} = \frac{14Ml^2}{\frac{7}{4}l \times 4M} = 2l$$

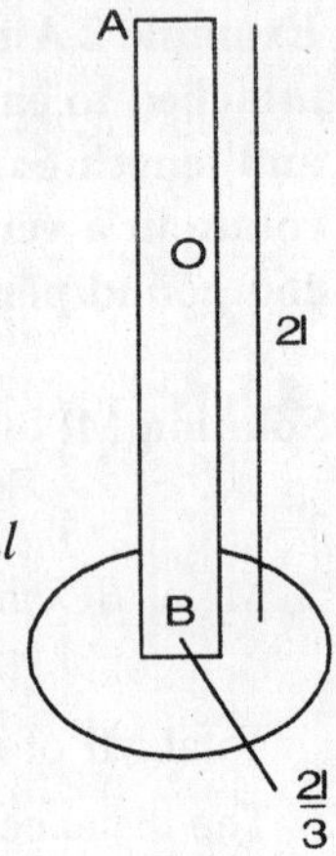

Fig. 7.14

$$\text{Time period } T = 2\pi\sqrt{\frac{L}{g}} = 2\pi\sqrt{\frac{2l}{g}}$$

Example 4 A uniform rod of mass *m* and length 2*a* is free to rotate about the end point A. A particle of mass *4m/3* can be attached at any point on the rod. Prove that the length of the SEP is minimum when the point of attachment is at a distance of *a*/2 from the point of suspension. Find max. length of SEP.

Solution MI of AB about *A* is $I_1 = \frac{1}{3}m(a)^2 + m(a)^2 = \frac{4}{3}ma^2$

MI of the particle at *P* about *A* is $I_2 = \frac{4}{3}mx^2$

$$\text{Total MI of the system} = I_1 + I_2 = \frac{4}{3}ma^2 + \frac{4}{3}mx^2 \qquad \ldots (1)$$

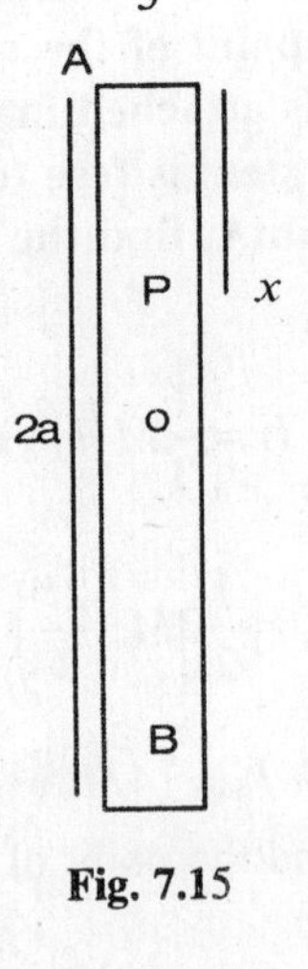

Fig. 7.15

The distance between CG and the point of suspension A is

$$h=\frac{m\times a+\frac{4}{3}m\times x}{m+\frac{4}{3}m}=\frac{3a+4x}{7} \qquad \dots (2)$$

In view of (1) and (2) we get,

$$\text{SEP length } l=\frac{\text{Total M.I. of the system}}{h\times\text{total mass of the system}}$$

$$=\frac{\frac{4}{3}ma^2+\frac{4}{3}mx^2}{\frac{3a+4x}{7}\times\frac{7m}{3}}=4\left(a^2+x^2\right)/(3a+4x)$$

For l to be minimum, we have

$$\frac{dl}{dx}=\frac{d}{dx}\left\{\frac{4\left(a^2+x^2\right)}{3a+4x}\right\}=0$$

$$\Rightarrow \quad 3ax+4x^2-2\left(a^2+x^2\right)=0$$

$$\Rightarrow \quad 2x^2+3ax-2a^2=0$$

$$\Rightarrow \quad x=\frac{a}{2},\ x=-2a \text{ (not possible)}$$

For $x=\frac{a}{2},\ \frac{d^2l}{dx^2}=50\frac{a^2}{(3a+2a)^3}=\frac{2}{5a}$ (positive)

Hence, when $x=\frac{a}{2}$, the length of SEP is minimum.

$$l_{\max}=\frac{4\left(a^2+\left(\frac{a}{2}\right)^2\right)}{3a+4\left(\frac{a}{2}\right)}=a$$

$$\text{Time period } T=2\pi\sqrt{\frac{l_{max}}{g}}=2\pi\sqrt{\frac{a}{g}}$$

Example 5 A uniform rod AB of mass M and length $2l$ is free to rotate about the end point A. A particle of mass m is attached at B. Find the length of the SEP and the time period.

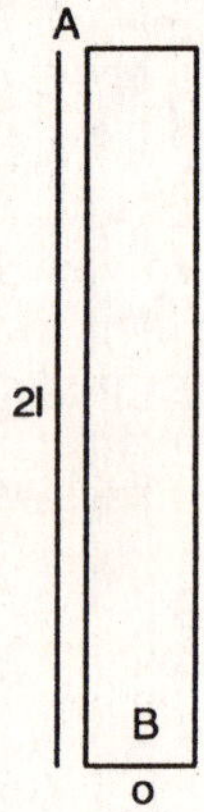

Fig. 7.16

Solution MI of AB about A is $I_1 = \frac{1}{3}M(l)^2 + M(l)^2 = \frac{4}{3}Ml^2$

MI of the particle at B about A is $I_2 = m(2l)^2$

Total MI of the system $= I_1 + I_2 = \frac{4}{3}Ml^2 + 4ml^2$ ... (1)

The distance between CG and the point of suspension A is

$$\Rightarrow \quad h = \frac{M \times l + m \times 2l}{M + m} = \frac{(2m+M)l}{M+m} \quad \text{... (2)}$$

In view of (1) and (2) we get,

$$\text{SEP length } l = \frac{\text{Total M.I. of the system}}{h \times \text{total mass of the system}}$$

$$= \frac{\frac{4}{3}Ml^2 + 4ml^2}{\frac{(2m+M)l}{M+m} \times (m+M)} = \frac{4(3m+M)l}{3(2m+M)}$$

$$\text{Time period } T = 2\pi\sqrt{\frac{l}{g}} = 2\pi\sqrt{\frac{4(3m+M)l}{3g(2m+M)}}.$$

REVIEW EXERCISE 7.1

1. A rod of length $2l$ and mass m and a particle of mass m_1 fixed at one end. The system oscillates as a compound pendulum about other end of the rod. Show that the period of small oscillation is

$$4\pi\sqrt{\frac{(m+3m_1)l}{3(m+2m_1)g}}.$$

2. A weightless rod ABC of length $2l$ is capable of free rotation about a horizontal axis through A. A particle of mass m is attached to B, the mid point of the rod, and another of mass m at C. Show that the length of SEP is $5l/3$.
3. Three equal particles are attached to a weightless rod at equal distances l apart. The system is suspended and is free to rotated about a point of the

rod distant x from the middle point. Find the period of small oscillation show that the time period is least when x = 0.82l approximately.

(**Hint.** $I = m(x-l)^2 + mx^2 + m(x+l)^2$, $h = x$, $T = 2\pi\sqrt{\frac{1}{g}\left(x+\frac{2l^2}{3x}\right)}$

is minimum when $\left(x+\frac{2l^2}{3x}\right)$ is minimum i.e. $x=\sqrt{\frac{2}{3}}l.$)

4. Show that the time period of a compound pendulum consists of solid sphere is minimum, if the axis is at a distance0. 63R approximately, from the centre of the sphere.

(**Hint.** $T = 2\pi\sqrt{\frac{1}{g}\left(x+\frac{2R^2}{5x}\right)}$ find minima of T when $x=\frac{2}{5}R$ = 0.63R).

5. A Uniform circular disc of radius r has a particle of mass equal to that of the disc fixed at a point on its circumference. The disc can turn freely about a fixed horizontal axis through its centre at right angleto its plane. Show that the length of SEP is 3r/2.

(Total MI $=\frac{1}{2}M(r)^2 + M(r)^2 = \frac{3}{2}Mr^2$, h = r/2).

8 Projectile Motion and Resisting Medium

Introduction

The knowledge of **Projectile motion and Resisting Medium** is of paramount importance in every sphere of life right from war to play ground. The knowledge of **Projectile motion and Resisting Medium** is immensely used in understanding the concept of external ballistic of gun, targeting of missile and from missile and the range of shell explosion. Thus in light of the importance of **Projectile motion and Resisting Medium**, we shall study the theory of **Projectile motion and Resisting Medium** and its consequences.

§ 1. Projectile and trajectory. The particle which is projected is known as **projectile** and the path traced out by the particle is known as its **trajectory**. The projection medium has negligible air resistance and constant gravity.

§ 2. *A particle of mass m is projected from a fixed point, into the air with velocity u in a direction making an angle with the horizontal. Discuss the motion of the particle.*

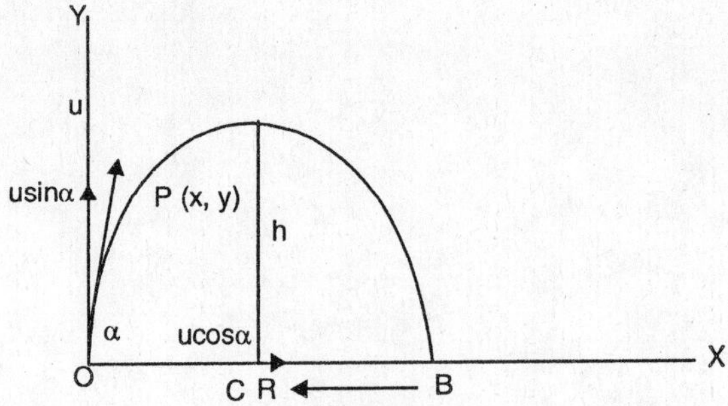

Fig. 8.1

Consider O, the point of projection, as origin. Let us consider, the vertical line through O as Y-axis, and the horizontal line through O and lying in the plane in which the particle being projected as X-axis.

Let particle be at P (x, y) after time t, then $\frac{d^2x}{dt^2}$ and $\frac{d^2y}{dt^2}$ will be the components of acceleration of the particle in the directions of X and Y respectively, also in sense of x and y increasing. The only force acting on the particle at P (x, y) its weight mg, acting vertically downwards.

Then the equation motion of particle in the horizontal direction is given by

$$m\frac{d^2x}{dt^2} = 0 \qquad \text{(i)}$$

Integrating (i) we get,

$u_x = \frac{dx}{dt}$ = constant, initially i.e. at O, the horizontal component of velocity = u cosα

$$\Rightarrow \quad u_x = \frac{dx}{dt} = u\cos\alpha \qquad \text{(ii)}$$

Hence horizontal component of velocity will remain constant throughout the motion.

The equation motion of particle in the vertical direction is given by

$$m\frac{d^2y}{dt^2} = -mg \qquad \text{(iii)}$$

Integrating (iii) we get, $u_y = \frac{dy}{dt} = -gt + C$, initially i.e. at O, the vertical component of velocity $= u$ sinα

$$\Rightarrow \quad u_y = \frac{dy}{dt} = u\sin\alpha - gt \qquad \text{(iv)}$$

Here u_x and u_y are horizontal and vertical components of velocity at P (x, y).

From (ii), on integrating we get,

$x = u\cos\alpha t + C'$, C′ is the constant of integration; initially when $t = 0$, then $x = 0$

$$\therefore \quad 0 = (u\cos\alpha)\times 0 + C' \Rightarrow C' = 0$$

$$\Rightarrow \quad x = u\cos\alpha t \qquad \text{(v)}$$

From (iv), on integrating we get, $y = u\sin\alpha t - \frac{1}{2}gt^2 + C''$,

C″ is the constant of integration; initially when $t = 0$, then $y = 0$

$$\therefore \quad 0 = (u\sin\alpha)\times 0 - \frac{1}{2}g\times 0 + C'' \Rightarrow C'' = 0$$

$$\Rightarrow \quad y = u\sin\alpha t - \frac{1}{2}gt^2 \tag{vi}$$

(v) and (vi) give us the position of a particle at any instant t i.e. coordinates of P (x, y) (in terms of t).

The equation of the path traced out by the particle is obtained by eliminating the variable t between (v) and (vi).

From (v) we have $t = \dfrac{x}{u\cos\alpha}$

Now putting the value of t in (vi) we get,

$$y = u\sin\alpha\left(\frac{x}{u\cos\alpha}\right) - \frac{1}{2}g\left(\frac{x}{u\cos\alpha}\right)^2$$

$$\Rightarrow \quad y = x\tan\alpha - \frac{1}{2}g\frac{x^2}{u^2\cos^2\alpha} \tag{vii}$$

the required equation of trajectory of the particle.

§ 3. *A particle of mass m is projected from a fixed point, into the air with velocity u in a direction making an angle α with the horizontal. Find the latus rectum, vertex, directrix and focus of the parabola traced out by the projectile.*

From Art 2 equation (vii) the equation of the path of projectile is given by

$$y = x\tan\alpha - \frac{1}{2}g\frac{x^2}{u^2\cos^2\alpha}$$

Now making perfect square to reduce the equation into standard parabola $X^2 = 4bY$ with the consideration that the coefficient of x^2 is one.

$$\Rightarrow \quad \frac{2u^2\cos^2\alpha}{g}y = \left(\frac{2u^2\cos^2\alpha}{g}\tan\alpha\right)x - x^2$$

Adding both sides $\left(\dfrac{u^2\cos\alpha\sin\alpha}{g}\right)^2$ for making RHS perfect square,

$$\Rightarrow \quad \left(\frac{u^2\cos\alpha\sin\alpha}{g}\right)^2 - \frac{2u^2\cos^2\alpha}{g}y$$

$$= \left(\frac{u^2\cos\alpha\sin\alpha}{g}\right)^2 - \frac{2u^2\cos\alpha}{g}\sin\alpha x + x^2$$

$$\left(x - \frac{u^2\cos\alpha\sin\alpha}{g}\right)^2 = -\frac{2u^2\cos^2\alpha}{g}\left(y - \frac{u^2\sin^2\alpha}{2g}\right) \quad \text{(i)}$$

Shift the origin to (x', y'), such that

$$x' = \frac{u^2\cos\alpha\,\sin\alpha}{g} \text{ and } y' = \frac{u^2\sin^2\alpha}{2g} \quad \text{(ii)}$$

Then (i) reduces into

$$X^2 = -4bY$$

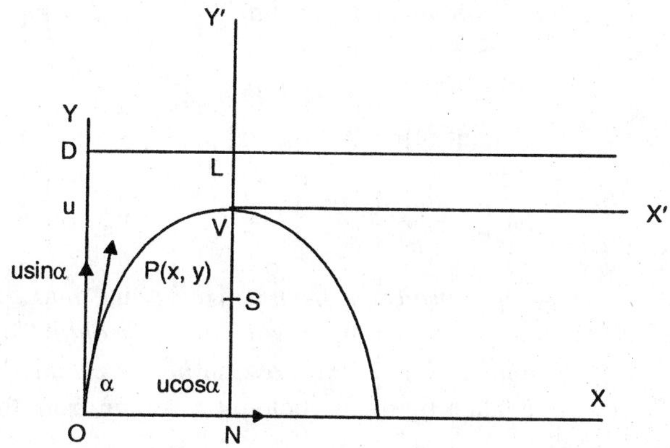

Fig. 8.2

Vertex of the parabola V is given by

$$(x', y') = \left(\frac{u^2\cos\alpha\sin\alpha}{g}, \frac{u^2\sin^2\alpha}{2g}\right) \quad \text{(ii)}$$

Latus rectum of the parabola

$$4b = \frac{2u^2\cos^2\alpha}{g} = \frac{2}{g}\text{(horizontal component of velocity)}^2 \quad \text{(iii)}$$

Directrix of the parabola is given by

$$Y = b \Rightarrow \left(y - \frac{u^2 \sin^2 \alpha}{2g} \right) = \frac{u^2 \cos^2 \alpha}{2g}$$

$\Rightarrow \quad y = \frac{u^2}{2g}$ (iv) required equation of directrix.

Again the abscissa or the x coordinate of the focus and vertex is same (see fig). However, the ordinate of the focus $S = VN - VS$, thus we have

$$\text{Ordinate of the focus } S = \frac{u^2 \sin^2 \alpha}{2g} - \frac{u^2 \cos^2 \alpha}{2g} = -\frac{u^2 \cos 2\alpha}{2g}$$

∴ Coordinates of the focus are

$$S\left(\frac{u^2 \cos\alpha \sin\alpha}{g}, -\frac{u^2 \cos 2\alpha}{2g} \right) \qquad \text{(v)}$$

§ 4. **Velocity at any point of the projectile.** *The velocity v at any point on the trajectory of the projectile is the resultant of the horizontal and vertical components of the velocity. In view of (ii) and (iv) of Art 2, we get.*

$$v = \sqrt{u_x^2 + u_y^2} = \sqrt{(u\cos\alpha)^2 + (u\sin\alpha - gt)^2}$$

The direction of the velocity from horizontal axis is given by

$$\theta = \tan^{-1}\left(\frac{u_y}{u_x} \right) = \tan^{-1}\left(\frac{u\sin\alpha - gt}{u\cos\alpha} \right)$$

§ 5. *A particle of mass m is projected from a fixed point, into the air with velocity u in a direction making an angle with the horizontal. Find (a) the time of flight (b) horizontal range (c) the greatest height attained.*

(a) **Time of flight.** Time taken by the particle in travelling from the point of projection O

To B, the point where the horizontal line through O intersects the path of the projectile, is known as time of flight; it is represented by T.

Vertical distance travelled by the particle in time t is given by

$$y = u\sin\alpha t - \frac{1}{2}gt^2 \qquad \text{(i)}$$

(u sin α is the initial velocity component in vertical direction)

At B, $y = 0$ and $t = T$ = time of flight. Equation (i) reduces into

$$0 = u\sin\alpha T - \frac{1}{2}gT^2$$

$\Rightarrow$ Time of flight $= T = \dfrac{2u\sin\alpha}{g} = \dfrac{2}{g} \times$ initial vertical velocity (ii)

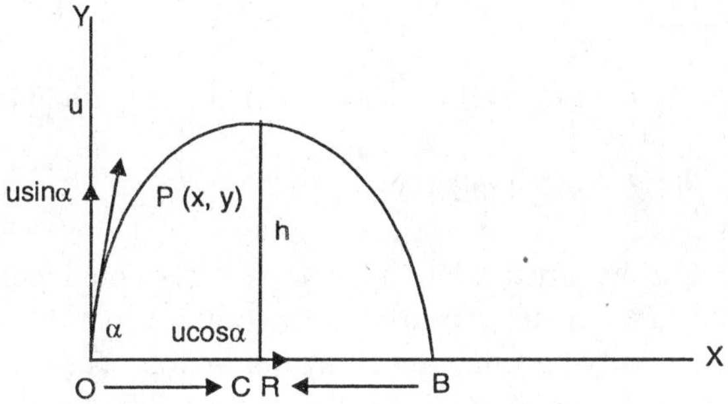

Fig. 8.3

(b) Horizontal range. Horizontal distance covered by the particle in time T (the time of flight) from the point of projection O, is known as Range of projectile and denoted by R. From figure OB = R.

$\Rightarrow$ $R = u\cos\alpha \times T$ (iii)

Now putting the value of T from (ii) we get,

$$R = u\cos\alpha \times \frac{2u\sin\alpha}{g}$$

$$= \frac{2}{g} \times \text{initial vertical velocity} \times \text{initial horizontal velocity}$$

$$R = u\cos\alpha \times \frac{2u\sin\alpha}{g} = \frac{u^2}{g}\sin 2\alpha = \frac{1}{2}$$

(abscissa of the vertex of the parabola) (iv)

Maximum horizontal range. From (iv) it is obvious that the range will be maximum if

$$\sin 2\alpha = 1 \Rightarrow \alpha = \frac{\pi}{4}$$

Hence range will be maximum if the particle is projected at an angle $\frac{\pi}{4}$ to the horizontal and is given by

$$R_{\max} = \frac{u^2}{g}$$

Direction of the range. We know that

$$R=\frac{u^2}{g}\sin 2\alpha=\frac{u^2}{g}\sin(\pi-2\alpha)=\frac{u^2}{g}\sin 2\left(\frac{\pi}{2}-\alpha\right)=\frac{u^2}{g}\sin 2\beta,$$

where $\beta=\left(\frac{\pi}{2}-\alpha\right)$

Thus we can say the range remains same, when the particle is projected at angle (α) or $\left(\frac{\pi}{2}-\alpha\right)$ with the X-axis.

(c) Greatest height attained. Since vertical velocity component is retarded or accelerated by gravity, thereby during the upward motion the stage will come where its velocity becomes zero. Let the time be T_1

$$0=u\sin\alpha-gT_1$$

$$\Rightarrow \quad T_1=\frac{u\sin\alpha}{g} \tag{v}$$

Vertical distance travelled by the particle in time T_1 is given by

$$CH=h=u\sin\alpha T_1-\frac{1}{2}gT_1^2$$

Now putting the value of T_1 from (v) we get,

$$h=u\sin\alpha\frac{u\sin\alpha}{g}-\frac{1}{2}g\left(\frac{u\sin\alpha}{g}\right)^2$$

$$\Rightarrow \quad h=\frac{u^2\sin^2\alpha}{2g}=\text{coordinate of the vertex of the parabola} \tag{vi}$$

Important results to be used

1. $u_x=\frac{dx}{dt}=u\cos\alpha(\text{constant}),\ u_y=\frac{dy}{dt}=u\sin\alpha-gt,$
2. $y=x\tan\alpha-\frac{1}{2}g\frac{x^2}{u^2\cos^2\alpha}$ (equation of trajectory of the particle parabola)
3. Vertex of the parabola V $\left(\frac{u^2\cos\alpha\sin\alpha}{g},\frac{u^2\sin^2\alpha}{2g}\right)$,

   Latus rectum $=\frac{2u^2\cos^2\alpha}{g}$,

Focus $S\left(\frac{u^2\cos\alpha\sin\alpha}{g}, -\frac{u^2\cos 2\alpha}{2g}\right)$ (equation of directrix)

4. Time of flight $= T = \frac{2u\sin\alpha}{g}$, $R = \frac{u^2}{g}\sin 2\alpha$, $R_{max} = \frac{u^2}{g}$,

when $\alpha = \frac{\pi}{4}$, $h = \frac{u^2\sin^2\alpha}{2g}$.

5. Velocity at any point $v = \sqrt{(u\cos\alpha)^2 + (u\sin\alpha - gt)^2}$

ILLUSTRATIVE EXAMPLES

Example 1 A particle is projected with a velocity *u* so that its range on a horizontal plane is twice the greatest height attained. Show that the range is $\frac{4u^2}{5g}$.

Solution Let α be the angle of projection.

According to question

$$R = 2h$$

$$\Rightarrow \quad R = \frac{u^2}{g}\sin 2\alpha = \frac{u^2}{g} 2\sin\alpha\cos\alpha = 2\frac{u^2\sin^2\alpha}{2g}$$

$$\Rightarrow \quad 2\sin\alpha\cos\alpha = \sin^2\alpha$$

$$\Rightarrow \quad \tan\alpha = 2$$

$$\Rightarrow \quad \sin\alpha = \frac{2}{\sqrt{5}}, \cos\alpha = \frac{1}{\sqrt{5}}$$

$$R = \frac{u^2}{g} 2\sin\alpha\cos\alpha = \frac{2u^2}{g}\frac{2}{\sqrt{5}}\frac{1}{\sqrt{5}} = \frac{4u^2}{5g}$$

Example 2 A bullet is aimed at a mark which is in a horizontal plane through the point of projection falls *a* feet short of it when the elevation is α and goes *b* feet far when it is projected with angle β. Show that if the velocity of projection be same in all cases, the proper elevation is $\frac{1}{2}\sin^{-1}\left[\frac{a\sin 2\beta + b\sin 2\alpha}{a+b}\right]$.

Solution Let O be the point of projection of the bullet

Which is aimed at C such that OC = c

Let ϕ be the required elevation to hit the point C

And u is the velocity of projection which is same in all cases.

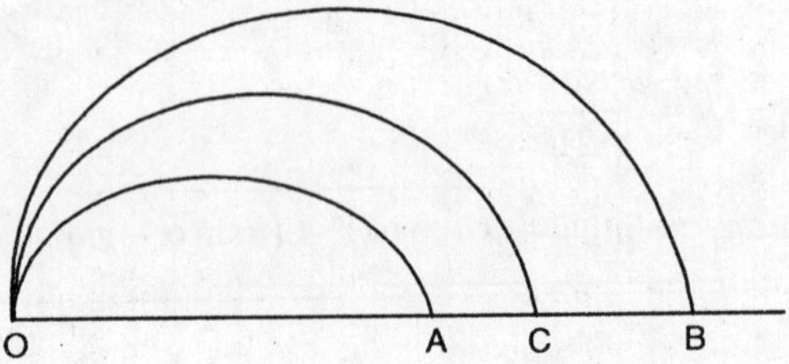

Fig. 8.4

When angle of projection is α then the range is OA = $c - a$

$$c - a = \frac{u^2}{g} \sin 2\alpha \qquad \text{... (1)}$$

When angle of projection is β then the range is OB = $c + b$

$$c + b = \frac{u^2}{g} \sin 2\beta \qquad \text{... (2)}$$

When angle of projection is ϕ then the range is OC = c

$$c = \frac{u^2}{g} \sin 2\phi \qquad \text{... (3)}$$

Multiplying (1) by b and (ii) by a and adding we get,

$$c(a+b) = \frac{u^2}{g}(b \sin 2\alpha + a \sin 2\beta) \qquad \text{... (4)}$$

Substituting the value of c from (3) in (4) we get,

$$\frac{u^2}{g} \sin 2\phi (a+b) = \frac{u^2}{g}(b \sin 2\alpha + a \sin 2\beta)$$

$$\Rightarrow \quad \sin 2\phi = \left[\frac{a \sin 2\beta + b \sin 2\alpha}{a+b}\right]$$

$$\Rightarrow \quad \phi = \frac{1}{2} \sin^{-1}\left[\frac{a \sin 2\beta + b \sin 2\alpha}{a+b}\right]$$

Example 3 Two bullets are fired from the same point with the same velocity but in different directions. If range in each case be R and the time of flight be T and T', prove that $R = \frac{1}{2} gTT'$.

Solution Let O be the point of projection of the bullets With velocity of projection u.

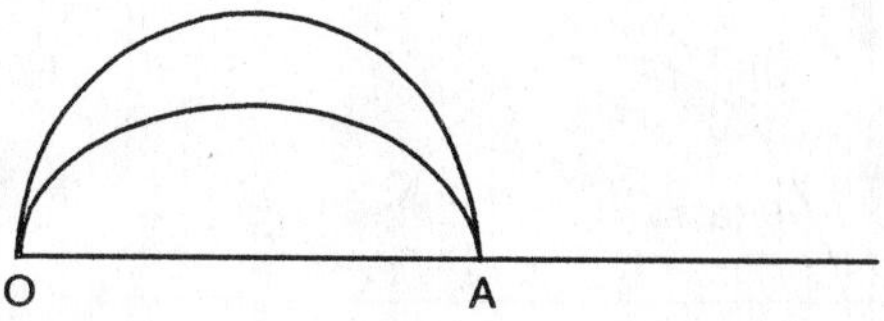

Fig. 8.5

When angle of projection is α then the range is O A

$$R = \frac{u^2}{g} \sin 2\alpha, \text{ Time of flight} = T = \frac{2u \sin \alpha}{g} \qquad \dots (1)$$

Then for the other bullet the velocity of projection is same and for the same range, the angle of projection should be $\beta = \left(\frac{\pi}{2} - \alpha\right)$, time of flight is given by

$$\text{Time of flight} = T' = \frac{2u \sin\left(\frac{\pi}{2} - \alpha\right)}{g} = \frac{2u \cos \alpha}{g} \qquad \dots (2)$$

From (1) and (2) we get,

$$\frac{1}{2} gTT' = \frac{1}{2} g \frac{2u \sin \alpha}{g} \frac{2u \cos \alpha}{g} = 2u \cos \alpha \times \frac{u \sin \alpha}{g} = \frac{u^2}{g} \sin 2\alpha = R$$

Example 4 A particle is projected from a point on the level of the ground, and its height is h when it is at horizontal distances a and $2a$, from its point of projection. Prove that the velocity of projection u is given by $u^2 = \frac{g}{4}\left(\frac{4a^2}{h} + 9h\right)$.

Solution Let α be the angle of projection and O be the point of projection. Let P and Q be two points on trajectory having same height h. Then according to question the coordinates of the points will be

P (a, h) and Q $(2a, h)$.

The equation of the path of projectile is given by

$$y = x\tan\alpha - \frac{1}{2}g\frac{x^2}{u^2\cos^2\alpha} \quad \text{... (1)}$$

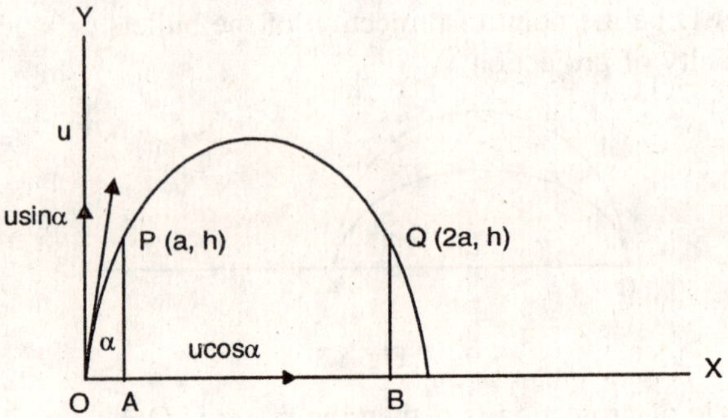

Fig. 8.6

As P (a, h) and Q $(2a, h)$, lie on (1), so we have,

$$h = a\tan\alpha - \frac{1}{2}g\frac{a^2}{u^2\cos^2\alpha} \quad \text{... (2)}$$

$$h = 2a\tan\alpha - \frac{1}{2}g\frac{4a^2}{u^2\cos^2\alpha} \quad \text{... (3)}$$

Subtracting (2) from (3)

$$0 = a\tan\alpha - \frac{3}{2}g\frac{a^2}{u^2\cos^2\alpha}$$

$$\Rightarrow \quad a\tan\alpha = \frac{3}{2}g\frac{a^2}{u^2\cos^2\alpha} \quad \text{... (4)}$$

In view of (2) and (4) we get,

$$h = \frac{3}{2}g\frac{a^2}{u^2\cos^2\alpha} - \frac{1}{2}g\frac{a^2}{u^2\cos^2\alpha} = g\frac{a^2}{u^2\cos^2\alpha}$$

$$\Rightarrow \quad \sec^2\alpha = \frac{u^2h}{a^2g} \quad \text{... (5)}$$

From (4) and (5) we have,

$$\sec^2\alpha = \frac{u^2h}{a^2g} = 1 + \tan^2\alpha = 1 + \left(\frac{3}{2}g\frac{a}{u^2}\frac{u^2h}{a^2g}\right)^2 \Rightarrow u^2 = \frac{g}{4}\left(\frac{4a^2}{h} + 9h\right)$$

Example 5 Shots are fired simultaneously from top and bottom of a vertical cliff of height h with elevations of α and β respectively strike an enemy bunker simultaneously. Show that the horizontal distance of the object from the cliff is given by $h/(\tan\beta - \tan\alpha)$.

Solution Let OO′ = h, be the height of cliff and P be the object. Let u and u' be the velocities of the shot fired from the top and bottom of the cliff respectively.

Since the shots are fired simultaneously and targeted the object P simultaneously ⇒ Time taken by each shot in reaching P is the same.

Let this time be t, then the horizontal component of the velocity remains constant throughout the motion and the horizontal distance travelled is due to horizontal component of velocity, we have

$$a = u\cos\alpha \times t = u'\cos\beta \times t$$

$$\Rightarrow \quad u\cos\alpha = u'\cos\beta \qquad \dots (1)$$

Also referred to horizontal and vertical lines through the points of projection as coordinate axes. The paths traced out by the shots projected from O and O′ are

$$y = x\tan\alpha - \frac{gx^2}{2u^2\cos^2\alpha} \qquad \dots (2)$$

And

$$y = x\tan\beta - \frac{gx^2}{2u'^2\cos^2\beta} \qquad \dots (3)$$

Let height of P above O be y', then the depth of P below O′ is $h-y'$.

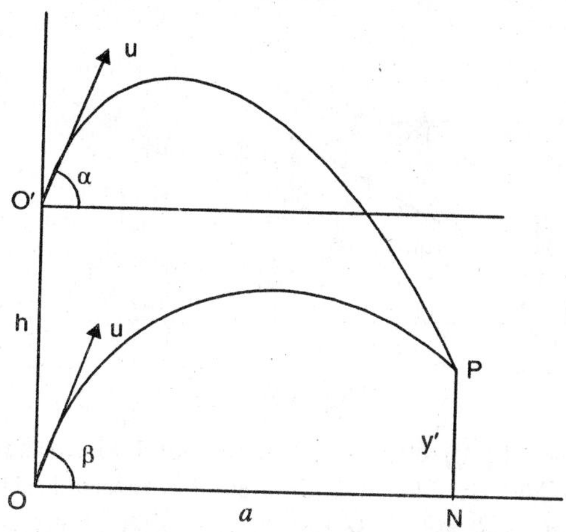

Fig. 8.7

Therefore, the coordinates of P w.r.t. to axes through O are (a, y') (and through O′ $(a, -(h-y'))$. In view of (2) and (3) we have

$$-(h-y') = a\tan\alpha - \frac{ga^2}{2u^2\cos^2\alpha} \qquad \dots (4)$$

$$y' = a\tan\beta - \frac{ga^2}{2u'^2\cos^2\beta} \qquad \dots (5)$$

(5)-(4) we get,

$$h = a(\tan\beta - \tan\alpha) - ga^2\left(\frac{1}{2u'^2\cos^2\beta} - \frac{1}{2u^2\cos^2\alpha}\right) \qquad \dots (6)$$

In view of (1) and (6) we have

$$a = h/(\tan\beta - \tan\alpha)$$

§ 6. Range and time of flight up the inclined plane

A particle of is projected from a fixed point, into the air with velocity u in a direction making an angle α with the horizontal on an inclined plane, inclined at an angle β to the horizontal. If the particle strikes the inclined plane Find (a) range in inclined plane (b) time of flight up the plane.

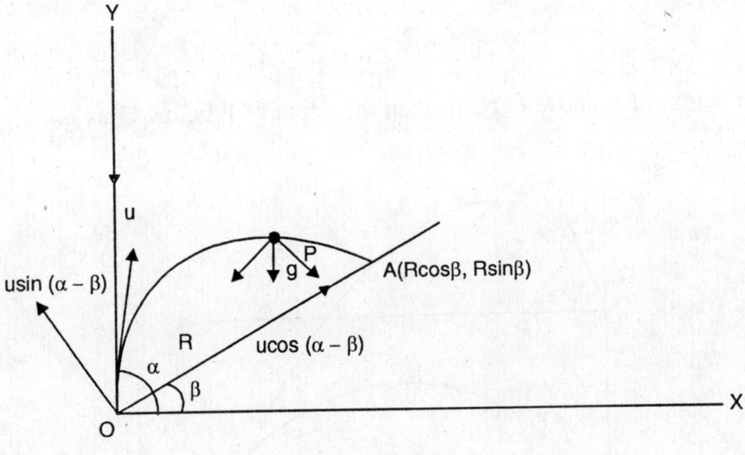

Fig. 8.8

Let O be the point of projection and A be the point where the particle strike the plane OA. Let OA = R. Then coordinates of A referred to horizontal and vertical line through O as axes (see fig.) are $(R\cos\beta, R\sin\beta)$.

The path traced out by the projectile is

$$y = x\tan\alpha - \frac{gx^2}{2u^2\cos^2\alpha} \quad \text{(i)}$$

$\therefore$ $A(R\cos\beta, R\sin\beta)$ lies on trajectory (i), thus we have

$$R\sin\beta = R\cos\beta\tan\alpha - \frac{g(R\cos\beta)^2}{2u^2\cos^2\alpha}$$

$$\Rightarrow \quad -\sin\beta + \cos\beta\tan\alpha = \frac{gR\cos^2\beta}{2u^2\cos^2\alpha}$$

$$\Rightarrow \quad (\cos\beta\sin\alpha - \cos\alpha\sin\beta)/\cos\alpha = \frac{gR\cos^2\beta}{2u^2\cos^2\alpha}$$

$$\Rightarrow \quad R = \frac{2u^2\cos\alpha\sin(\alpha-\beta)}{g\cos^2\beta} \quad \text{(ii)}$$

Equation (ii) gives the range on an inclined plane (up the inclined plane) Maximum range up the inclined plane. From Equation (ii) we have,

$$R = \frac{2u^2\cos\alpha\sin(\alpha-\beta)}{g\cos^2\beta} = \frac{u^2}{g\cos^2\beta}\left[\sin(2\alpha-\beta) - \sin\beta\right] \quad \text{(iii)}$$

If u and β are constants, then from (iii), R is maximum if $\sin(2\alpha-\beta)$ is maximum i.e.

$$\sin(2\alpha-\beta) = 1 \Rightarrow 2\alpha - \beta = \pi/2$$

$$\Rightarrow \quad \alpha = \frac{\pi}{4} + \frac{\beta}{2} \quad \text{(iv)}$$

$$R_{max} = \frac{u^2}{g\cos^2\beta}[1-\sin\beta] = \frac{u^2}{g(1-\sin^2\beta)}[1-\sin\beta]$$

$$\Rightarrow \quad R_{max} = \frac{u^2}{g(1+\sin\beta)} \quad \text{(v)}$$

Also from (iv) we have

$$\alpha - \beta = \alpha - \left(2\alpha - \frac{\pi}{2}\right) = \frac{\pi}{2} - \alpha \quad \text{(vi)}$$

From (vi) we can say that the direction of projection bisects the angle between the vertical line at O and the line of greatest slope.

Time of flight

The component of velocity of projection at right angle to the inclined plane OA is $u\sin(\alpha-\beta)$.

Further from figure it is clear that the component of acceleration working on the particle along and perpendicular to the plane are $g\sin\beta$ and $g\cos\beta$ respectively.

If T be the required time of flight from O to A, then by taking the motion perpendicular to the plane

$$0 = u\sin(\alpha-\beta)T - \frac{1}{2}g\cos\beta T^2 \quad \left(\text{Using } s = ut + \frac{1}{2}ft^2\right)$$

$$\Rightarrow \quad T = 2u\sin(\alpha-\beta)/g\cos\beta \tag{vii}$$

Equation (vii) gives the Time of flight on an inclined plane (up the inclined plane).

§ 6. Range and time of flight down the inclined plane

Let O be the point of projection and u be the velocity of projection and α the angle which u make with horizontal. Let the inclined plane make angle β with the horizontal plane.

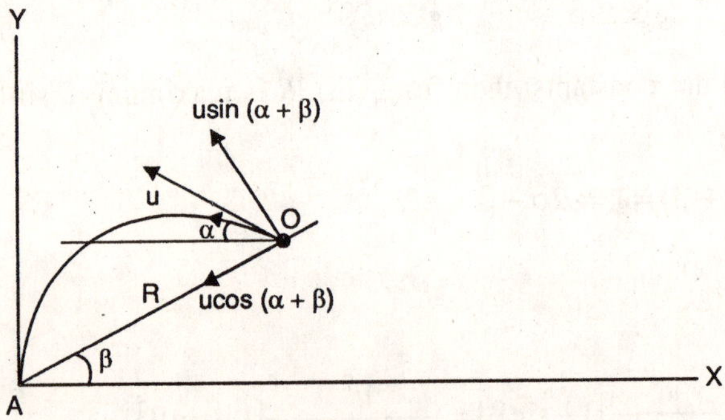

Fig. 8.9

Time of flight

The component of velocity of projection at right angle to the inclined plane OA is $u\sin(\alpha+\beta)$.

Further from figure it is clear that the component of acceleration working on the particle along and perpendicular to the plane are $g\sin\beta$ and $g\cos\beta$ respectively.

If T be the required time of flight from O to A, then by taking the motion perpendicular to the plane

$$0 = u\sin(\alpha+\beta)T - \frac{1}{2}g\cos\beta T^2 \left(\text{Using } s = ut + \frac{1}{2}ft^2\right)$$

$$\Rightarrow \quad T = 2u\sin(\alpha+\beta)/g\cos\beta \qquad \text{(i)}$$

Equation (i) gives the Time of flight on an inclined plane (down the inclined plane).

Range down the inclined plane

The component of velocity of projection at right angle to the inclined plane OA is $u\sin(\alpha+\beta)$.

Further from figure it is clear that the component of acceleration working on the particle along and perpendicular to the plane are $g\sin\beta$ and $g\cos\beta$ respectively.

If T be the required time of flight from O to A, then by taking the motion along the plane

$$R = u\cos(\alpha+\beta)T + \frac{1}{2}g\sin\beta T^2 \qquad \text{(ii)} \left(\text{Using } s = ut + \frac{1}{2}ft^2\right)$$

Substituting the value of T from (i) in (ii) we get

$$R = u\cos(\alpha+\beta)\frac{2u\sin(\alpha+\beta)}{g\cos\beta} + \frac{1}{2}g\sin\beta\left(\frac{2u\sin(\alpha+\beta)}{g\cos\beta}\right)^2$$

$$\Rightarrow \quad R = 2u^2\frac{\sin(\alpha+\beta)}{g\cos^2\beta}\left[\cos(\alpha+\beta)\cos\beta + \sin(\alpha+\beta)\sin\beta\right]$$

$$\Rightarrow \quad R = \frac{2u^2\cos\alpha\sin(\alpha+\beta)}{g\cos^2\beta} \qquad \text{(iii)}$$

Equation (iii) gives the range on an inclined plane (up the inclined plane) Maximum range up the inclined plane. From Equation (iii) we have,

$$R = \frac{2u^2\cos\alpha\sin(\alpha-\beta)}{g\cos^2\beta} = \frac{u^2}{g\cos^2\beta}\left[\sin(2\alpha+\beta)+\sin\beta\right] \qquad \text{(iv)}$$

If u and β are constants, then from (iv), R is maximum if $\sin(2\alpha+\beta)$ is maximum i.e.

$$\sin(2\alpha+\beta) = 1 \Rightarrow 2\alpha+\beta = \pi/2$$

$$\Rightarrow \quad \alpha = \frac{\pi}{4} - \frac{\beta}{2} \tag{v}$$

$$R_{max} = \frac{2u^2}{g\cos^2\beta}[1+\sin\beta] = \frac{2u^2}{g(1-\sin^2\beta)}[1+\sin\beta]$$

$$\Rightarrow \quad R_{max} = \frac{u^2}{g(1-\sin\beta)} \tag{vi}$$

Note: *Replace* β *by* $-\beta$ *in motion up the plane we get the results of down the plane.*

Important results to be used
Projected up the inclined plane
$R = \dfrac{2u^2\cos\alpha\sin(\alpha-\beta)}{g\cos^2\beta}$, $R_{max} = \dfrac{u^2}{g(1+\sin\beta)}$, $T = 2u\sin(\alpha-\beta)/g\cos\beta$
Projected down the inclined plane
$R = \dfrac{2u^2\cos\alpha\sin(\alpha+\beta)}{g\cos^2\beta}$, $R_{max} = \dfrac{u^2}{g(1-\sin\beta)}$, $T = 2u\sin(\alpha+\beta)/g\cos\beta$

ILLUSTRATIVE EXAMPLES

Example 1 For a given velocity of projection the maximum range down an inclined plane is three times the range up the inclined plane, show that the inclination of the plane is 30°.

Solution Let u be the velocity of projection and the inclined plane make angle β with the horizontal plane

According to question

$$R_{max} \text{ down the plane} = 3R_{max} \text{ up the plane}$$

$$\Rightarrow \quad \frac{u^2}{g(1-\sin\beta)} = 3\frac{u^2}{g(1+\sin\beta)} \Rightarrow (1+\sin\beta) = 3(1-\sin\beta)$$

$$\Rightarrow \quad 4\sin\beta = 2 \Rightarrow \beta = 30°.$$

Example 2 If a gun be placed on an inclined plane, the maximum range in the horizontal direction is harmonic mean of maximum ranges up and down the plane.

Solution Let u be the velocity of projection and the inclined plane make angle β with the horizontal plane

$$R_{max} \text{ up the plane} = \frac{u^2}{g(1+\sin\beta)} = R$$

$$R_{max} \text{ down the plane} = \frac{u^2}{g(1-\sin\beta)} = R'$$

Harmonic mean of R and R'

$$= \frac{2RR'}{R+R'} = \frac{\left\{\dfrac{2u^4}{g(1-\sin^2\beta)}\right\}}{\dfrac{u^2}{g(1+\sin\beta)} + \dfrac{u^2}{g(1-\sin\beta)}} = \frac{u^2}{g} = \text{Max horizontal range.}$$

Example 3 The angular elevation of an enemy position on a hill of *h* metres high is β. Show that in order to shell it, the initial velocity of projection must not be less than $\sqrt{[gh(1+\text{cosec}\,\beta)]}$.

Solution Let and the inclined plane make angle β **with the horizontal**

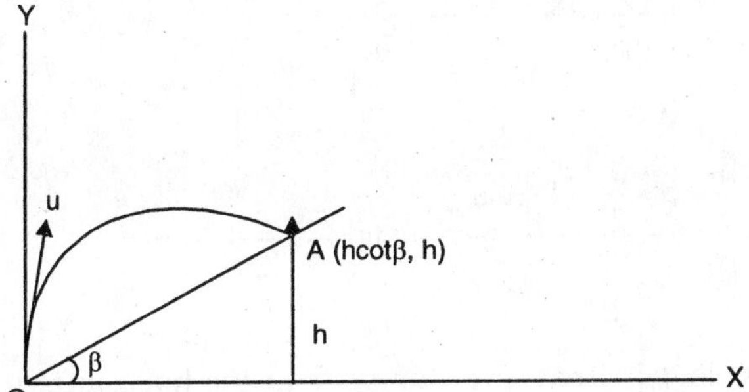

Fig. 8.10

Let O be the point of projection of the shell and A be the position of the enemy at a height h above O the plane OA then, OA = R = $h\,\text{cosec}\,\beta$. If u be the minimum initial velocity of projection to shell the enemy, OA must be maximum range up the inclined plane.

$$R_{max} \text{ up the plane} = \frac{u^2}{g(1+\sin\beta)} = h\,\text{cosec}\,\beta$$

$$\Rightarrow \quad u^2 = g(1+\sin\beta)h\operatorname{cosec}\beta$$

$$\Rightarrow \quad u = \sqrt{[gh(1+\operatorname{cosec}\beta)]}.$$

Example 4 The line joining A to B is inclined at an angle β to the horizon. Show that the least velocity required to shoot from A to B is $\tan\left(\frac{\pi}{4}+\frac{\beta}{2}\right)$ times the least velocity required to shoot from B to A.

Solution Let the least velocities required to shoot from A to B and shoot from B to A be u and v respectively. The inclined plane makes angle β with the horizontal plane

$$R_{max} \text{ up the plane} = \frac{u^2}{g(1+\sin\beta)} = R \qquad \ldots (1)$$

$$R_{max} \text{ down the plane} = \frac{v^2}{g(1-\sin\beta)} = R \qquad \ldots (2)$$

From (1) and (2) we have,

$$u^2/v^2 = \frac{1+\sin\beta}{1-\sin\beta} = \frac{\left(\sin\frac{\beta}{2}+\cos\frac{\beta}{2}\right)^2}{\left(\sin\frac{\beta}{2}-\cos\frac{\beta}{2}\right)^2}$$

$$\frac{u}{v} = \frac{\sin\frac{\beta}{2}+\cos\frac{\beta}{2}}{\sin\frac{\beta}{2}-\cos\frac{\beta}{2}} = \frac{1+\tan\frac{b}{2}}{1-\tan\frac{\beta}{2}} = \tan\left(\frac{\pi}{4}+\frac{\beta}{2}\right)$$

Example 5 A shot is fired at an angle α to the horizontal up a hill of inclination β to the horizontal. Show that it strikes the hill horizontally if $\tan\alpha = 2\tan\beta$.

Solution If T be the time of flight, then

$$\Rightarrow \quad T = 2u\sin(\alpha-\beta)/g\cos\beta \qquad \ldots (1)$$

Since the shot strikes the hill horizontally after time T from its start, so it vertical component of velocity should vanish after time T.

$$0 = u\sin\alpha - gT \qquad \ldots (2) \text{ (Using } v = u + ft\text{)}$$

$$T = \frac{u\sin\alpha}{g} \qquad \ldots (3)$$

From (1) and (3) we get

$$2u\sin(\alpha-\beta)/g\cos\beta = \frac{u\sin\alpha}{g}$$

$$2\sin(\alpha-\beta) = \sin\alpha\cos\beta$$

$$\Rightarrow \quad 2\sin\alpha\cos\beta = \cos\alpha\sin\beta$$

Dividing by cos β, we get,

$$\tan\alpha = 2\tan\beta$$

REVIEW EXERCISE 8.1

1. If R be the range of a projectile on a horizontal plane and h is the maximum height for a given angle of projection, show that the maximum horizontal range with same velocity of projection is $2h+\frac{R^2}{h}$.

   $\left(\textbf{Hint: } \text{Find } 2h+\frac{R^2}{h}=\frac{u^2}{g}=R_{max}\right)$

2. An aeroplane is moving horizontally with velocity360 km/hr drops a bomb from a height of 1102.5 meters. How far should it be horizontally from the object to be hit.
3. If H_1 and H_2 are the greatest height reached by a particle in two projectile paths of same horizontal range R, then prove that $R=4\sqrt{H_1H_2}$.
   Hint. α and $90°-\alpha$ (are two angles of projection for same R).
4. Shots are fired simultaneously from top and bottom of a vertical cliff with elevations of α and β respectively strike an enemy bunker simultaneously. Show that if a be the horizontal distance of the object from the cliff the height of the cliff is $a\,(\tan\beta-\tan\alpha)$.
5. An aeroplane is flying with constant velocity v and at a constant height h. Show that, if gun is fired point blank at the aeroplane after it has passed directly over the gun and when angle of elevation as seen from the gun is α, the shell will hit the areoplane provided $2(u\cos\alpha-v)\tan^2\alpha=gh$. Here u is the velocity of the shell.
6. A shot is fired at an angle α to the horizontal up a hill of inclination β to the horizontal. Show that it strikes the hill normally if tanα = cotβ + 2 tanβ. (**Hint.** It component of velocity along the plane should vanish after time T, $0=u\sin(\alpha-\beta)-g\sin\beta T$, and T is the time of flight).
7. A particle is projected with velocity u from a point on a plane inclined at an angle β to the horizontal. If R and R' be the maximum ranges up and

down the inclined plane, prove that $\frac{1}{R}+\frac{1}{R'}$ is independent of the inclination of the plane.

§ 7. Resisting Medium

Due to the presence of air or any other fluid in the medium, there is some resistance to the motion of the particle. This resistance usually varies either with velocity or with some power of velocity.

(a) **Descent:** *If a particle is falling under gravity in a medium having resistance to the motion known as Descent.*

(b) **Ascent:** *If a particle is projected upwards in a medium having resistance to the motion known as Ascent.*

Resistance varies as the velocity

(a) **Descent:** Find the motion of the particle, if a particle is falling under gravity in a medium having resistance varies as the velocity.
The net force acting on the particle is given by $t = 0$, $x = 0$, $v = 0$

$$mf = mg - mR$$

$$\Rightarrow \quad f = g - R \tag{i}$$

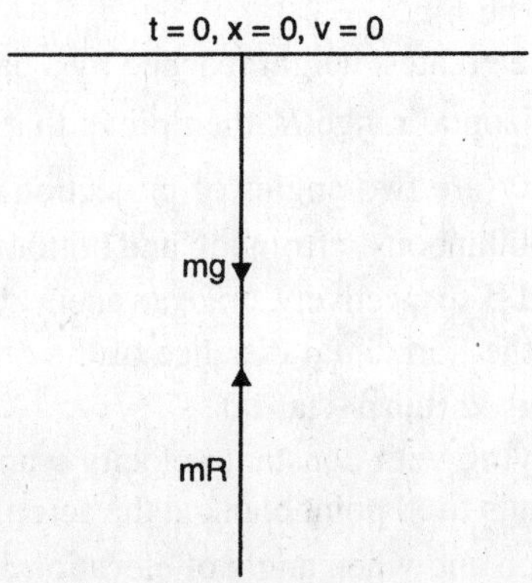

Fig. 8.11

But it is given that resistance is directly proportional to Velocity

$$R \propto v \Rightarrow R = kv \tag{ii}$$

$$f = g - kv$$

$$\Rightarrow \quad \frac{d^2x}{dt^2} = g\left[1 - \frac{1}{V}\frac{dx}{dt}\right] \tag{iii}$$

Here g and k both are constant so that $V = g/k$ be a constant is known as terminal velocity.

If $\dfrac{dx}{dt} = V,$ then from (iii) it is clear that acceleration of the particle would be zero, then motion will be unresisted and particle will fall with the constant velocity V.

Rewriting (iii) as

$$\frac{d^2x}{dt^2} + \frac{g}{V}\frac{dx}{dt} = g \tag{iv}$$

Now integrating (iv) w.r.t. t we get,

$$\frac{dx}{dt} + \frac{g}{V}x = gt + C$$

Initially $x = 0 = v$ or $\dfrac{dx}{dt} \Rightarrow C = 0$

$$\frac{dx}{dt} + \frac{g}{V}x = gt \tag{v}$$

Equation (v) is linear differential equation, so it solution is given by

$$xe^{\frac{g}{V}t} = \int gte^{\frac{g}{V}t}\,dt + C'$$

Solving RHS, integration by parts we get,

$$xe^{\frac{g}{V}t} = \left(Vt - \frac{V^2}{g}\right)e^{\frac{g}{V}t} + C' \tag{vi}$$

Initially $x = 0 = t \Rightarrow C' = \left(\dfrac{V^2}{g}\right)$

Now equation (vi) reduces to

$$xe^{\frac{g}{V}t} = \left(Vt - \frac{V^2}{g}\right)e^{\frac{g}{V}t} + \frac{V^2}{g}$$

$$\Rightarrow \quad x = Vt - \frac{V^2}{g} + \frac{V^2}{g}e^{-\frac{g}{V}t} \tag{vii}$$

Here equation (vii) represents the distance covered by the particle at any instant t.

Now differentiating (vii) with respect to t we get

$$v=\frac{dx}{dt}=V\left(1-e^{-\frac{g}{V}t}\right) \tag{viii}$$

Here equation (vii) represents velocity of the particle at any instant t.

Note. *Above results can also be derived by considering equation (vi) as linear Differential Equation for* $v=\frac{dx}{dt}$.

$$\frac{dv}{dt}=v\frac{dv}{dx}=g-\frac{g}{V}v \tag{ix}$$

Now applying variable separable method we get

$$\frac{v}{V-v}dv=\frac{g}{V}dx \tag{x}$$

On integrating (x) we get

$$-V\log\left(V-v\right)=\frac{g}{V}x+C'' \tag{xi}$$

Initially $x=0=v$ or $\Rightarrow C''=-V\log\left(V\right)$

Now equation (xi) reduces to

$$x=\frac{V^2}{g}\log\frac{v}{\left(V-v\right)}-\frac{vV}{g} \tag{xii}$$

Here equation (xii) represents the distance covered by the particle in terms of velocity v and terminal velocity V.

(b) **Ascent:** Find the motion of the particle, if a particle is projected upwards with the velocity U in a medium having resistance varies as the velocity.

The net force acting on the particle is given by

$$mf=-mg-mR$$

$$\Rightarrow \quad f=-g-R \tag{i}$$

But it is given that resistance is directly proportional to Velocity

$$R\propto v\Rightarrow R=kv \tag{ii}$$

$$f=-g-kv$$

$$\Rightarrow \quad \frac{d^2x}{dt^2}=-g\left[1+\frac{1}{V}\frac{dx}{dt}\right] \tag{iii}$$

Here g and k both are constant so that $V = g/k$ be a constant is known as terminal velocity

Rewriting (iii) as

$$\frac{d^2x}{dt^2}+\frac{g}{V}\frac{dx}{dt}=-g \tag{iv}$$

Now integrating (iv) w.r. t. t we get,

$$\frac{dx}{dt}+\frac{g}{V}x=-gt+C$$

Initially $t=0=x, \frac{dx}{dt}=U \Rightarrow C=U$

$$\frac{dx}{dt}+\frac{g}{V}x=-gt+U \tag{v}$$

Equation (v) is linear differential equation, so it solution is given by

$$xe^{\frac{g}{V}t}=\int(-gt+U)e^{\frac{g}{V}t}\,dt+C'$$

Solving RHS, integration by parts we get,

$$xe^{\frac{g}{V}t}=\left\{-\left(Vt-\frac{V^2}{g}\right)+\frac{UV}{g}\right\}e^{\frac{g}{V}t}+C' \tag{vi}$$

Initially $x=0=t \Rightarrow C'=-\left(\frac{V^2}{g}+\frac{UV}{g}\right)$

Now equation (vi) reduces to

$$xe^{\frac{g}{V}t}=\left\{-\left(Vt-\frac{V^2}{g}\right)+\frac{UV}{g}\right\}e^{\frac{g}{V}t}+-\left(\frac{V^2}{g}+\frac{UV}{g}\right)$$

$$\Rightarrow \quad x=\left\{-\left(Vt-\frac{V^2}{g}\right)+\frac{UV}{g}\right\}+-\left(\frac{V^2}{g}+\frac{UV}{g}\right)e^{\frac{-g}{V}t} \tag{vii}$$

Here equation (vii) represents the distance covered by the particle at any instant t.

Now differentiating (vii) with respect to t we get

$$v=\frac{dx}{dt}=(U+V)e^{\frac{-g}{V}t}-V \tag{viii}$$

At highest point $v=\frac{dx}{dt}=0$

$$\Rightarrow \quad \frac{g}{V}t = log\frac{U+V}{V} \tag{ix}$$

In view of (vii) and (ix) we get the greatest height attended by particle

$$h = \left\{-\left(\frac{V^2}{g}\log\frac{U+V}{V} - \frac{V^2}{g}\right) + \frac{UV}{g}\right\} + -\left(\frac{V^2}{g} + \frac{UV}{g}\right)\frac{V}{U+V}$$

$$h = \frac{UV}{g} - \frac{V^2}{g}\log\frac{U+V}{V} \tag{x}$$

Here equation (x) represents highest point attended by particle.

Note. *Above results can also be derived by considering equation (vi) as linear Differential Equation for* $v = \frac{dx}{dt}$.

$$\frac{dv}{dt} = v\frac{dv}{dx} = -g - \frac{g}{V}v \tag{xi}$$

Now applying variable separable method we get

$$\frac{v}{V+v}dv = -\frac{g}{V}dx \tag{xii}$$

On integrating (xii) we get

$$v - V\log(V+v) = -\frac{g}{V}x + C'' \tag{xi}$$

Initially $x = 0, v = U$ or $\Rightarrow C'' = U - V\log(U+V)$

Now equation (xi) reduces to

$$x = \frac{V^2}{g}\log\frac{V+v}{(V+U)} + \frac{V(U-v)}{g} \tag{xii}$$

At highest point $v = 0$, then equation (xii) can be reduced to (x) highest point achieved by the particle.

If = λV, then in view of (x), the greatest height attended by the particle is given by

$$h = \frac{V^2}{g}(\lambda - \log(1+\lambda))$$

REVIEW EXERCISE 8.2

1. If a particle is falling under gravity in a medium having resistance varies as the square of the velocity. Prove that (a) $v^2 = V^2\left[1 - e^{-\frac{2g}{V^2}x}\right]$,

(b) $v = V \tan\frac{gt}{V}$, (c) $x = \frac{V^2}{g}\left[\log\cosh\frac{gt}{V}\right]$ where $V = \sqrt{\frac{g}{k}}$ is the terminal velocity. (**Hint.** $R = kv^2$).

2. If a particle is projected upwards with the velocity U in a medium having resistance varies as the square of the velocity. Prove that greatest height reached by the particle is $h = \frac{V^2}{g}\left[\log\sec\frac{gt}{V}\right]$, where $V = \sqrt{\frac{g}{k}}$ is the terminal velocity. (**Hint.** $R = kv^2$ and at highest point velocity is zero).
3. A particle is projected with velocity V along the smooth horizontal plane in a medium whose resistance per unit mass is k (velocity). Show that the velocity v after time t and distance x in that time are given by $v = Ve^{-kt}$, $x = \frac{V}{k}\left(1 - e^{-kt}\right)$. (**Hint.** $f = -kv$)

Appendix

APPLICATION OF DYNAMICS IN DEFENCE

Introduction. Dynamics is widely used in defence field, like the effect of shell explosion, the range of shell explosion, ballistic of gun or missile, targeting of missile from air and the effective range of air bombing. In the appendix section we will discuss the brief application of dynamics in defence as an application of previously studied chapters.

TYPE 1

Shell explosion problem. Such problems are dealt with the help of conservation of energy and impact theory.

Example 1 A shell of mass m is projected from a gun of mass *M* by an explosion which generates kinetic energy *E*. Prove that the initial velocity of the shell is $\sqrt{\{2EM/M(m+M)\}}$, with the assumption that at the instant of explosion the gun is free to recoil.

Solution Let u and v be the velocities of the shell and recoil of gun respectively. Then by LCM we have,

Momentum of a shell = momentum of a gun

$$mu = Mv \qquad \text{... (1)}$$

Also from energy

$$\Rightarrow \qquad E = \frac{1}{2}mu^2 + \frac{1}{2}Mv^2 \qquad \text{... (2)}$$

Eliminating v from (1) and (2), we get

$$E = \frac{1}{2}mu^2 + \frac{1}{2}M\left(\frac{mu}{M}\right)^2 = \frac{m(m+M)}{2M}u^2$$

$$\Rightarrow \quad u = \sqrt{\{2EM/M(m+M)\}}$$

Example 2 A shell is moving with velocity *U*, explodes into two fragments of mass *m* and *m*′, which move with velocities *u* and *u*′ in opposite directions. Show that the gain of kinetic energy is $\frac{1}{2}\frac{mm'}{(m+m')}(u+u')^2$.

Solution Initial mass of the shell is $(m + m')$, Thus the gain of kinetic energy by explosion is E.

Gain of KE = KE after explosion – KE before explosion

$$E=\left(\frac{1}{2}mu^2+\frac{1}{2}m'u'^2\right)-\left(\frac{1}{2}(m+m')U^2\right) \qquad \ldots (1)$$

By conservation of the momentum

$$(m+m')U=mu-m'u' \Rightarrow U=\frac{mu-m'u'}{m+m'} \qquad \ldots (2)$$

Substituting U from (2) in (1) we get

$$E=\frac{1}{2}\frac{m'}{(m+m')}(u+u')^2$$

Example 3 A shell is at rest, an internal explosion breaks it into two fragments of mass m and m', which move in a straight line and opposite direction. If x is the distance between them after a time t, show that the work done by the explosion is $\frac{1}{2}\frac{mm'}{(m+m')}\left(\frac{x}{t}\right)^2$.

Solution Initial mass of the shell is $(m+m')$ which move with velocities u and u' in a straight line and opposite direction, by conservation of the momentum

$$(m+m')\times 0=mu-m'u' \Rightarrow mu=m'u' \qquad \ldots (1)$$

Distances after time t of the fragments m and m'

$x_1=ut$ and $x_2=u't$ in opposite direction, thus we have

$$x=x_1+x_2=(u+u')t \qquad \ldots (2)$$

Solving (1) and (2) we get,

$$u=\frac{m'}{(m+m')}\frac{x}{t} \text{ and } u'=\frac{m}{(m+m')}\frac{x}{t} \qquad \ldots (3)$$

Work done by the explosion

$$W=\left(\frac{1}{2}mu^2+\frac{1}{2}m'u'^2\right) \qquad \ldots (4)$$

Substituting the values of u and u' from (3) in (4) we get

$$W=\frac{1}{2}\frac{mm'}{(m+m')}\left(\frac{x}{t}\right)^2$$

Example 4 A shell of mass $m+m'$ is fired with a given velocity in a given direction. At the highest point of its path, the shell explodes into two fragments of mass m and m'. The explosion produces an additional kinetic energy E and fragments separate in a horizontal directions. Show that

the fragments strike the ground at A and B, then $AB = \frac{V}{g}\sqrt{2E\left(\frac{1}{m}+\frac{1}{m'}\right)}$, where V is the vertical component of velocity.

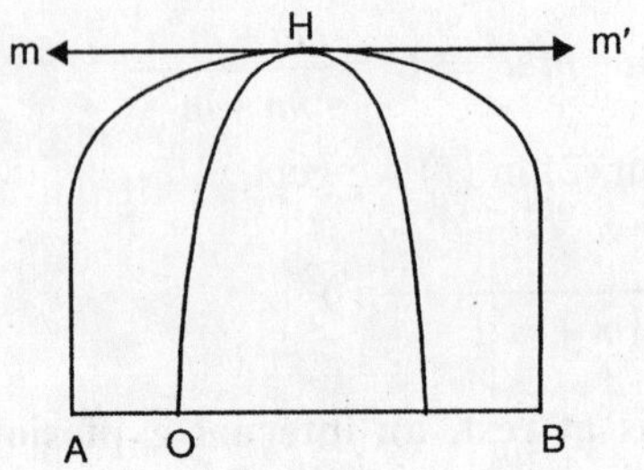

Fig. E.4

Solution. Let O be a point from which the shell is fired and H be the highest point at which the explosion takes place. Let U be the horizontal component of initial velocity of shell and T is the time taken by the shell to reach at H. u and u' speed of m and m' respectively.

KE of $m + m'$ just before explosion is $\frac{1}{2}(m+m')U^2$

The generation of kinetic energy by explosion is EKE just after the explosion is $\frac{1}{2}(m+m')\,U^2+E$

By conservation of energy

$$\frac{1}{2}(m+m')U^2+E=\frac{1}{2}mu^2+\frac{1}{2}m'u'^2 \qquad \ldots (1)$$

Then by LCM, we have

$$(m+m')U = mu - m'u' \qquad \ldots (2)$$

Eliminating U from (1) and (2) we get

$$u+u'=\sqrt{2E\left(\frac{m+m'}{mm'}\right)} \qquad \ldots (3)$$

Considering the vertical motion of $m + m'$ from O to H

$$0 = V - gT \Rightarrow T = V/g \qquad \ldots (4)$$

The fragments m and m' will take the same time T to reach the ground. During this motion the relative velocity of the fragments is $u + u'$. So we have

$$AB=(u+u')T=\frac{V}{g}\sqrt{2E\left(\frac{m+m'}{mm'}\right)}=\frac{V}{g}\sqrt{2E\left(\frac{1}{m}+\frac{1}{m'}\right)}$$

Example 5 A shell is moving with velocity U in the line AB. An internal explosion which generates an energy E, explodes into two fragments of mass m and m', which move in the line AB. Show that their velocities are $U+\sqrt{\{2Em'/m(m+m')\}}$ and $U-\sqrt{\{2Em/m'(m+m')\}}$.

Solution. Let the velocities of the fragment of masses m and m' be v and v' after explosion respectively.

Then by LCM, we have

$$(m+m')U = mv + m'v' \qquad \ldots (1)$$

By conservation of energy

$$\frac{1}{2}(m+m')U^2 + E = \frac{1}{2}mv^2 + \frac{1}{2}m'v'^2$$

$$\Rightarrow \quad (m+m')U^2 + 2E = mv^2 + m'v'^2 \qquad \ldots (2)$$

Also, practically, we find that $v = U + x/m$ and $v' = U - x/m'$. These values are also satisfying (1) for any value of x thus, making the statement mathematically true. These values must also satisfy (2) then we have,

$$(m+m')U^2 + 2E = m(U+x/m)^2 + m'(U-x/m')^2$$

$$\Rightarrow \quad (m+m')U^2 + 2E = (m+m')U^2 + x^2\left(\frac{1}{m}+\frac{1}{m'}\right)$$

$$\Rightarrow \quad x^2 = \{2Emm'/(m+m')\} \Rightarrow x = \sqrt{\{2Emm'/(m+m')\}} \qquad \ldots (3)$$

Hence

$$v = U + x/m = U + \sqrt{\{2Em'/m(m+m')\}}$$

$$v' = U - x/m = U - \sqrt{\{2Em'/m(m+m')\}}$$

Example 6 A shell is moving with velocity U in the line AB. An internal explosion which generates an energy E, explodes into two fragments of mass m and m', which move in the same line. Show that their relative velocity after explosion will be $\sqrt{\{2E(m+m')/mm'\}}$.

Solution. Let the velocities of the fragment of masses m and m' be v and v' after explosion respectively.

Then by LCM, we have

$$(m+m')U = mv + m'v' \qquad \ldots (1)$$

By conservation of energy

$$\frac{1}{2}(m+m')U^2+E=\frac{1}{2}mv^2+\frac{1}{2}m'v'^2$$

$$\Rightarrow \quad (m+m')U^2+2E=mv^2+m'v'^2 \qquad \text{... (2)}$$

On squaring (1), we get,

$$(m+m')^2U^2=(mv+m'v')^2 \qquad \text{... (3)}$$

On multiplication of (2) by $(m+m')$ we get,

$$(m+m')^2U^2+2(m+m')E=(m+m')(mv^2+m'v'^2) \qquad \text{... (4)}$$

Substracting (3) from (4) we get,

$$2(m+m')E=(mm')(v-v')^2$$

$$\Rightarrow \quad (v-v')=\sqrt{\{2E(m+m')/mm'\}}$$

Relative velocity of two parts after explosion

$$=(v-v')=\sqrt{\{2E(m+m')/mm'\}}\cdot$$

TYPE 2

The targeting of missile or bomb from air or ground and the effective range. These problems are dealt with the help of projectile motion.

Example 7 An aeroplane is moving horizontally with velocity 360 km/hr drops a bomb from a height of 1102.5 meters. How far should it be horizontally from the object to be hit?

Solution Let O be the point of projection and B the object to be hit.

Initial horizontal velocity $u_x=\dfrac{360\times1000}{3600}$ =100 m/s, Initial vertical velocity $u_y=0$

Let t be the time by the bomb in moving from O to B

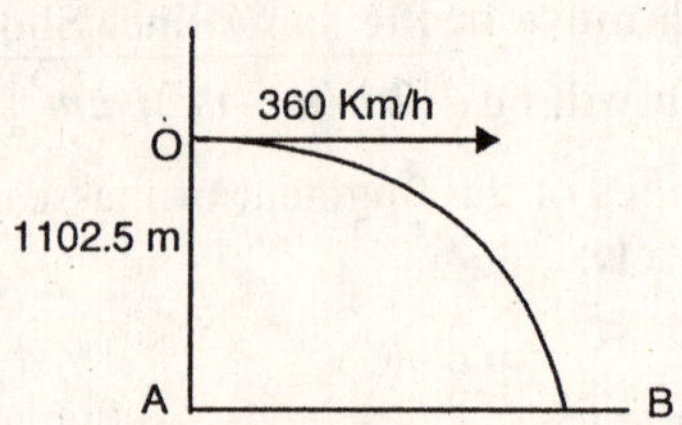

Fig. E.7

$$1102.5 = 0 + \frac{1}{2} \times 9.8 \times t^2 \text{ (vertical motion)}$$

$\Rightarrow \quad t = 15$ sec

Required distance AB = Horizontal velocity $\times t = 100 \times 15 = 1500$ metres.

Example 8 An aeroplane is moving horizontally with velocity 240 mile/hr drops a bomb from a height of 1200 yards. How far should it be horizontally from the object to be hit?

Solution Let O be the point of projection and B the object to be hit.

Initial horizontal velocity $u_x = \frac{240 \times 1760}{3600} = \frac{352}{3}$ yard/s = 352 ft/sec.

Initial vertical velocity $u_y = 0$

Let t be the time by the bomb in moving from O to B

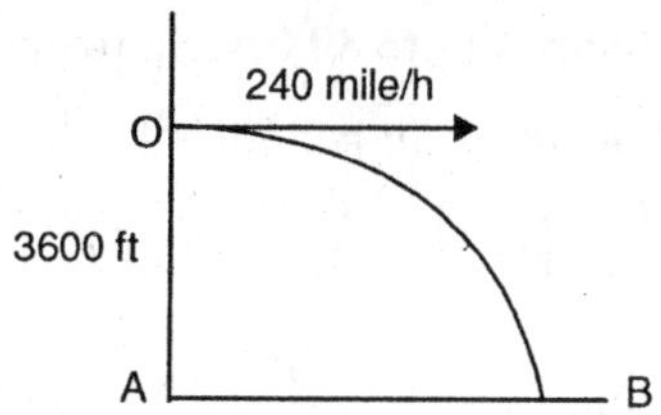

Fig. E.8

$$3600 = 0 + \frac{1}{2} \times g \times t^2 = \frac{1}{2} \times 32 \times t^2 \text{ (vertical motion) } (g = 32 \text{ ft/sec}^2)$$

$\Rightarrow \quad t = 15$ sec

Required distance AB = Horizontal velocity $\times t = 352 \times 15 = 5280$ ft = 1 mile.

Example 8 A shell bursts on contact with the ground and pieces from it fly in all direction with velocities upto 49 m/s. Show that the man 122.5 metre away is in danger for $5\sqrt{2}$ seconds.

Solution Given that Range R =122.5 mts, u = 49 m/s

Range of projectile is given by $R = \frac{u^2}{g} \sin 2\alpha \Rightarrow 122.5 = \frac{49 \times 49}{9.8} \sin 2\alpha$

$\Rightarrow \quad \sin 2\alpha = \frac{1}{2} \Rightarrow \alpha = 15°$

Since there are always two directions of projection α and $(90° - \alpha)$ for given range, hence for the range of 122.5 m there will be two directions of projection i.e. 15° and 90° – 15° = 75°.

Let T and T' be the time of flight of these two cases, then

$$T = \frac{2u\sin\alpha}{g} = \frac{2u}{g}\sin 15^\circ$$

$$T' = \frac{2u\sin(90-\alpha)}{g} = \frac{2u}{g}\sin 75^\circ$$

The man is in danger for a time

$$T' - T = \frac{2u}{g}(\sin 75^\circ - \sin 15^\circ) = \frac{2u}{g} 2\sin 45^\circ \sin 30^\circ$$

$$T' - T = \frac{4 \times 49 \times \frac{1}{\sqrt{2}} \times \frac{1}{2}}{9.8} = 5\sqrt{2} \text{ seconds.}$$

Example 9 A shell bursts on contact with the ground and pieces from it fly in all direction with velocities upto 80 ft/s. Show that the man 100 ft away is in danger for $\frac{5}{2}\sqrt{2}$ seconds. (g = 32 ft/s^2).

Solution Given that Range R = 100 ft, u = 80 ft/s

Range of projectile is given by $R = \frac{u^2}{g}\sin 2\alpha \Rightarrow 100 = \frac{80 \times 80}{32}\sin 2\alpha$

$$\Rightarrow \quad \sin 2\alpha = \frac{1}{2} \Rightarrow \alpha = 15^\circ$$

Since there are always two directions of projection α and (90° – α) for given range, hence for the range of 100 ft there will be two directions of projection i.e. 15° and 90° – 15° = 75°.

Let T and T' be the time of flight of these two cases, then

$$T = \frac{2u\sin\alpha}{g} = \frac{2u}{g}\sin 15^\circ$$

$$T' = \frac{2u\sin(90-\alpha)}{g} = \frac{2u}{g}\sin 75^\circ$$

The man is in danger for a time

$$T' - T = \frac{2u}{g}(\sin 75^\circ - \sin 15^\circ) = \frac{2u}{g} 2\sin 45^\circ \sin 30^\circ$$

$$T' - T = \frac{4 \times 80 \times \frac{1}{\sqrt{2}} \times \frac{1}{2}}{32} = \frac{5}{2}\sqrt{2} \text{ seconds.}$$

Example 10 Shots are fired simultaneously from top and bottom of a vertical cliff with elevations of α and β respectively strike an enemy bunker simultaneously. Show that if *a* be the horizontal distance of the object from the cliff the height of the cliff is $a(\tan\beta - \tan\alpha)$.

Solution Let $OO' = h$, be the height of cliff and P be the object. Let u and u' be the velocities of the shot fired from the top and bottom of the cliff respectively.

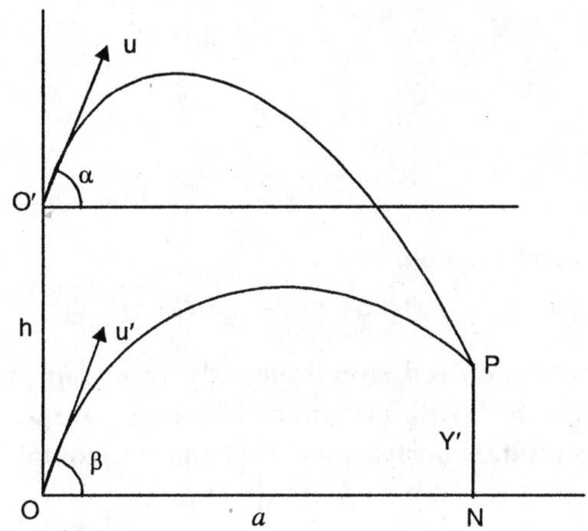

Fig. E.10

Since the shots are fired simultaneously and targeted the object P simultaneously ⇒ Time taken by each shot in reaching P is the same.

Let this time be t, then the horizontal component of the velocity remains constant throughout the motion and the horizontal distance travelled is due to horizontal component of velocity, we have

$$a = u\cos\alpha \times t = u'\cos\beta \times t$$

$$\Rightarrow \quad u\cos\alpha = u'\cos\beta \qquad \ldots (1)$$

Also referred to horizontal and vertical lines through the points of projection as coordinate axes. The paths traced out by the shots projected from O and O' are

$$y = x\tan\alpha - \frac{gx^2}{2u^2\cos^2\alpha} \qquad \ldots (2)$$

and

$$y = x\tan\beta - \frac{gx^2}{2u'^2\cos^2\beta} \qquad \ldots (3)$$

Let height of P above O be y', then the depth of P below O' is $h - y'$.

Therefore, the coordinates of P w.r.t. to axes through O are $(a - y')$ and through O' $(a, -(h - y'))$. In view of (2) and (3) we have

$$-(h-y') = a\tan\alpha - \frac{ga^2}{2u^2\cos^2\alpha} \qquad \text{... (4)}$$

$$y' = a\tan\beta - \frac{ga^2}{2u'^2\cos^2\beta} \qquad \text{... (5)}$$

(5)-(4) we get,

$$h = a(\tan\beta - \tan\alpha) - ga^2\left(\frac{1}{2u'^2\cos^2\beta} - \frac{1}{2u^2\cos^2\alpha}\right) \qquad \text{... (6)}$$

In view of (1) and (6) we have

$$h = a(\tan\beta - \tan\alpha)$$

Example 11 Shots are fired simultaneously from top and bottom of a vertical cliff of height h with elevations of α and β respectively strike an enemy bunker simultaneously. Show that the horizontal distance of the object from the cliff is given by $h/(\tan\beta - \tan\alpha)$.

Solution Let $OO' = h$, be the height of cliff and P be the object. Let u and u' be the velocities of the shot fired from the top and bottom of the cliff respectively.

Since the shots are fired simultaneously and targeted the object P simultaneously $\Rightarrow$ Time taken by each shot in reaching P is the same.

Let this time be t, then the horizontal component of the velocity remains constantthroughout the motion and the horizontal distance travelled is due to horizontal component of velocity, we have

$$a = u\cos\alpha \times t = u'\cos\beta \times t$$

$$\Rightarrow \quad u\cos\alpha = u'\cos\beta \qquad \text{... (1)}$$

Also referred to horizontal and vertical lines through the points of projection as coordinate axes. The paths traced out by the shots projected from O and O' are

$$y = x\tan\alpha - \frac{gx^2}{2u^2\cos^2\alpha} \qquad \text{... (2)}$$

and

$$y = x\tan\beta - \frac{gx^2}{2u'^2\cos^2\beta} \qquad \text{... (3)}$$

Let height of P above O be y', then the depth of P below O' is $h - y'$.

Therfore, the coordinates of P w.r.t. to axes through O are (a, y') and through O' $(a, -(h - y'))$. In view of (2) and (3) we have

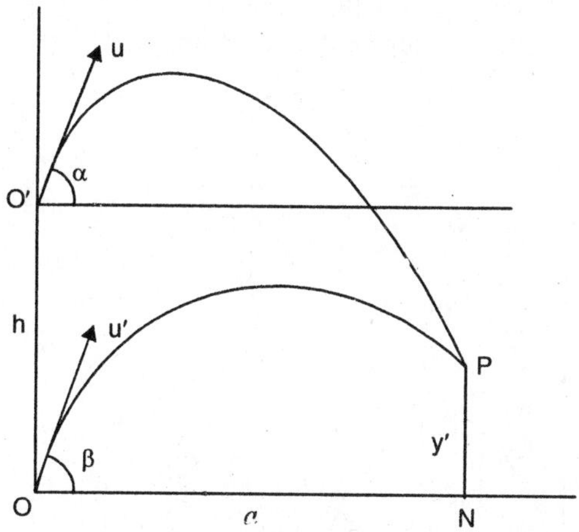

Fig. E.11

$$-(h-y') = a\tan\alpha - \frac{ga^2}{2u^2\cos^2\alpha} \quad \text{... (4)}$$

$$y' = a\tan\beta - \frac{ga^2}{2u'^2\cos^2\beta} \quad \text{... (5)}$$

(5)-(4) we get,

$$h = a(\tan\beta - \tan\alpha) - ga^2\left(\frac{1}{2u'^2\cos^2\beta} - \frac{1}{2u^2\cos^2\alpha}\right) \quad \text{... (6)}$$

In view of (1) and (6) we have

$$a = \frac{h}{(\tan\beta - \tan\alpha)}.$$

Targeting a moving object

Example 11 An aeroplane is flying with constant velocity v and at a constant height h. Show that, if gun is fired point blank at the aeroplane after it has passed directly over the gun and when angle of elevation as seen from the gun is α, the shell will hit the aeroplane provided $2(u\cos\alpha - v)\tan^2\alpha = gh$. Here u is the velocity of the shell.

Solution. Let O be the point of projection of the shell, A be the position of the aeroplane when the shot was fired. Fired point blank means the initial velocity u of the shell is along OA i.e. the angle of projection α.

The aeroplane is flying at a height h along the line BD with velocity v. C is the point where the shell can hit the aeroplane if they reach C at the same time.

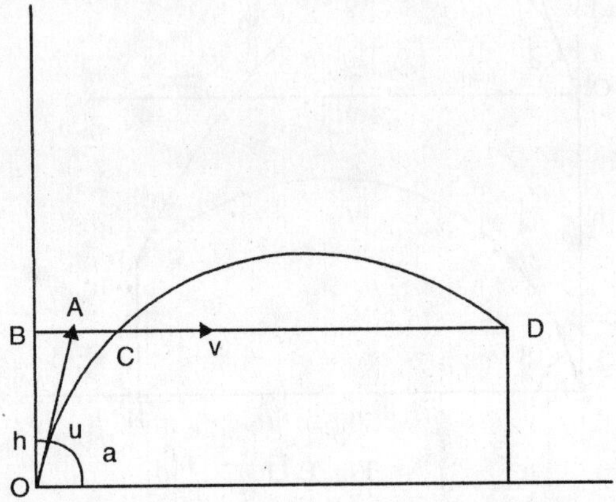

Fig. E.11

The starting point of the shell is O and same time the position of plane is at A. Let the shell hit the aeroplane after time t.

The distance moved by the aeroplane in time $t = AC = v.t$ ... (1)

Horizontal distance moved by the shell in time $t = BC = u\cos\alpha.t$ ... (2)

From triangle OBC, we have

$$BA = h\cot\alpha \quad \text{... (3)}$$

$$BC = BA + AC$$

$$\Rightarrow \quad u\cos\alpha.t = h\cot\alpha + v.t$$

$$t = \frac{h\cot\alpha}{(u\cos\alpha - v)} \quad \text{... (4)}$$

The vertical motion of the shell from O to C, and using (4)

$$h = u\sin\alpha.t - \frac{1}{2}gt^2 = \left\{u\sin\alpha - \frac{1}{2}g\left(\frac{h\cot\alpha}{(u\cos\alpha - v)}\right)\right\}\left(\frac{h\cot\alpha}{(u\cos\alpha - v)}\right)$$

$$\Rightarrow \quad 2h(u\cos\alpha - v)^2 = \left\{2u\sin\alpha(u\cos\alpha - v) - gh\cot\alpha\right\}h\cot\alpha$$

$$\Rightarrow \quad 2(u\cos\alpha - v)^2 - 2u\cos\alpha(u\cos\alpha - v) = gh\cot^2\alpha$$

$$\Rightarrow \quad 2(u\cos\alpha - v)\tan^2\alpha = gh.$$

Example 12 A gun fires a shell with a muzzle velocity *u*, show that the farthest horizontal distance at which an aeroplane at a height *h* can be hit is $\left(\frac{u}{g}\right)\sqrt{(u^2-2gh)}$ and then the elevation of gun is $\tan^{-1}\left\{\frac{u}{\sqrt{(u^2-2gh)}}\right\}$.

Solution Let O be the point of projection of the shell, A be the position of the aeroplane when the shot was fired with muzzle velocity u of the shell is along OA i.e. the angle of projection α.

The aeroplane is flying at a height h along the line BD with velocity v. C is the point where the shell can hit the aeroplane if they reach C at the same time.

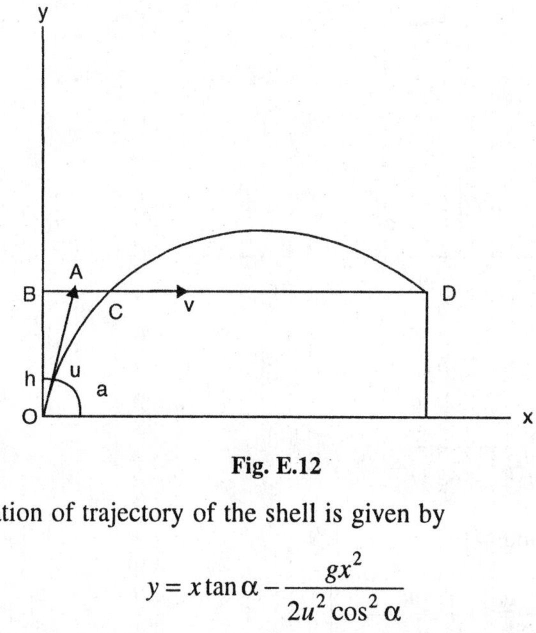

Fig. E.12

The equation of trajectory of the shell is given by

$$y = x\tan\alpha - \frac{gx^2}{2u^2\cos^2\alpha} \quad \dots (1)$$

If R is the horizontal distance of the aeroplane from the point of projection, then the coordinates of the aeroplane $C(R, h)$ should satisfy the equation of trajectory of the shell (1), if is hit by shell.

$$h = R\tan\alpha - \frac{gR^2\sec^2\alpha}{2u^2} \quad \dots (2)$$

If range R is maximum, then $\frac{dR}{d\alpha} = 0$ and $\frac{d^2R}{d\alpha^2}$ = negative

Differentiating both sides of (2) w.r.t. α, we get,

$$0 = R\sec^2\alpha + \tan\alpha\frac{dR}{d\alpha} - \frac{g}{2u^2}\left\{2R\sec^2\alpha\frac{dR}{d\alpha} + 2R^2\sec^2\alpha\tan\alpha\right\}$$

$$\Rightarrow \quad \frac{dR}{d\alpha}\left\{\tan\alpha - \frac{gR\sec^2\alpha}{u^2}\right\} = \frac{gR^2\sec^2\alpha\tan\alpha}{u^2} - R\sec^2\alpha$$

If $\frac{dR}{d\alpha} = 0$ then $\frac{gR^2\sec^2\alpha\tan\alpha}{u^2} - R\sec^2\alpha = 0$

$$\Rightarrow \quad \tan\alpha = \frac{u^2}{gR} \quad (3) \quad (\text{As } \sec^2\alpha \neq 0)$$

Putting the value of $\tan\alpha = \frac{u^2}{gR}$ in (2) we get,

$$h = R\frac{u^2}{gR} - \frac{gR^2\left[1 + \left(\frac{u^2}{gR}\right)^2\right]}{2u^2} = \frac{u^2}{2g} - \frac{gR^2}{2u^2}$$

$$\Rightarrow \quad R^2 = \frac{u^2}{g^2}(u^2 - 2gh)$$

$$\Rightarrow \quad R = \left(\frac{u}{g}\right)\sqrt{(u^2 - 2gh)}$$

In view of (3) and (4)

$$\tan\alpha = \frac{u^2}{gR} = \left\{\frac{u}{\sqrt{(u^2 - 2gh)}}\right\}$$

$$\alpha = \tan^{-1}\left\{\frac{u}{\sqrt{(u^2 - 2gh)}}\right\}$$

Important Formulae

IMPACT AND MOTION IN TWO DIMENSION IN POLAR COORDINATES

1. LCM $\Rightarrow$ $m_1u_1 + m_2u_2 = m_1v_1 + m_2v_2$ (Direct impact)

2. NEL $\Rightarrow$ $v_2 - v_1 = -e(u_2 - u_1)$

3. $v_1 = \dfrac{m_2(1+e)u_2 + (m_1 - em_2)u_1}{m_1 + m_2}$, $v_2 = \dfrac{m_1(1+e)u_1 + (m_2 - em_1)u_2}{m_1 + m_2}$

4. Pulse of blow on m_1 mass $= -\dfrac{m_1m_2\left[(u_1 - u_2)(1+e)\right]}{m_1 + m_2}$ pulse of blow on m_2 mass.

5. Change in kinetic energy (KE) $= -\dfrac{1}{2}\dfrac{m_1m_2}{m_1 + m_2}(1 - e^2)(u_2 - u_1)^2$

   $v_1 \cos\theta - v_2 \cos\phi = -e(u_1 \cos\alpha - u_2 \cos\beta)$ (NEL on Oblique Impact)

   $m_1v_1 \cos\theta + m_2v_2 \cos\phi = (m_1u_1 \cos\alpha + m_2u_2 \cos\beta)$ (LCM on Oblique Impact)

6. $v_1 \cos\theta = \dfrac{m_1u_1 \cos\alpha + m_2u_2 \cos\beta - em_2(u_1 \cos\alpha - u_2 \cos\beta)}{m_1 + m_2}$

7. $v_2 \cos\phi = \dfrac{m_1u_1 \cos\alpha + m_2u_2 \cos\beta + em_1(u_1 \cos\alpha - u_2 \cos\beta)}{m_1 + m_2}$

8. Change in kinetic energy (KE) $= -\dfrac{1}{2}\dfrac{m_1m_2(1 - e^2)(u_1 \cos\alpha - u_2 \cos\beta)^2}{m_1 + m_2}$

9. Radial velocity $(u) = \frac{dr}{dt} = \dot{r}$

10. Transverse velocity $(v) = r\frac{d\theta}{dt} = r\dot{\theta}$

11. Radial Acceleration $(f_r) = \frac{d^2r}{dt^2} - r\left(\frac{d\theta}{dt}\right)^2 = \ddot{r} - r\dot{\theta}^2$

12. Transverse Acceleration $(f_t) = 2\frac{dr}{dt}\frac{d\theta}{dt} + r\frac{d^2\theta}{dt^2} = r\ddot{\theta} + 2\dot{r}\dot{\theta} = \frac{1}{r}\frac{d(r^2\dot{\theta})}{dt}$

15. Resultant velocity $|\vec{v}| = \sqrt{(\dot{r})^2 + (\dot{r}\theta)^2}$.

16. $\frac{d}{dt}\hat{r} = \omega\hat{s},\ \frac{d}{dt}\hat{s} = -\omega\hat{s}$ where $\frac{d\theta}{dt} = \omega$.

17. $p = r\sin\phi$ and $\tan\phi = r\frac{d\theta}{dr}$.

18. $\frac{1}{p^2} = u^2 + \left(\frac{du}{d\theta}\right)^2$ where $u = \frac{1}{r}$

CENTRAL ORBIT

1. $r^2\dot{\theta} = h$ here h is a constant, Areal velocity $= \frac{h}{2}$.

2. $f = h^2u^2\left(\frac{d^2u}{d\theta^2} + u\right),\ f = \frac{h^2}{p^3}\frac{dp}{dr}$

3. $v^2 = \frac{h^2}{p^2} = h^2\left(u^2 + \left(\frac{du}{d\theta}\right)^2\right)$. Here $u = 1/r$.

(SHM + STRINGS)

1. $\frac{d^2x}{dt^2} = -\mu x$

2. $v = \frac{dx}{dt} = -\sqrt{\mu}\sqrt{(a^2 - x^2)}$.

3. $x = a\cos\left(\sqrt{\mu}t\right)$

4. $T = \frac{2\pi}{\sqrt{\mu}}, n = \frac{1}{T}$

5. $T = \frac{\lambda}{l_0}(l - l_0)$

6. Time period for horizontal elastic string $= 2\pi\sqrt{\frac{ma}{\lambda}}$

7. Time period for vertical elastic string $= 2\pi\sqrt{\frac{ma}{\lambda}} = 2\pi\sqrt{\frac{b}{g}}$

MI + COMPOUND PENDULAM

1. $I = Mk^2$
2. M.I of ring about any diameter $= \frac{1}{2}Ma^2$
3. Rod of length $2a$ and mass M M.I. about the line through the centre and perpendicular to the rod $= \frac{1}{3}Ma^2$
4. MI of disc about any diameter $= \frac{1}{4}Ma^2$, MI about the line through centre O and perpendicular to disc plane $= \frac{1}{2}Ma^2$.
5. The MI of the sphere about any diameter $= \frac{2}{5}Ma^2$. The MI of the spherical shell about any diameter $= \frac{2}{3}Ma^2$.
6. Time period of compound pendulum

$$T = 2\pi\sqrt{\frac{I_A}{Mgh}} = 2\pi\sqrt{\frac{I_G + Mh^2}{Mgh}} = 2\pi\sqrt{\frac{k^2 + h^2}{gh}}.$$

7. Length of simple equivalent pendulum(SEP),

$$l = \frac{I_G + Mh^2}{Mh} = \frac{I_A}{Mh} = \frac{k^2 + h^2}{h}.$$

Projectile Motion

1. $u_x = \dfrac{dx}{dt} = u\cos\alpha$ (constant), $u_y = \dfrac{dy}{dt} = u\sin\alpha - gt$,

2. $y = x\tan\alpha - \dfrac{1}{2}g\dfrac{x^2}{u^2\cos^2 a}$ (equation of trajectory of the particle parabola)

3. Vertex of the parabola $V\left(\dfrac{u^2\cos\alpha\sin\alpha}{g}, \dfrac{u^2\sin^2\alpha}{2g}\right)$, Latus rectum $= \dfrac{2u^2\cos^2\alpha}{g}$, Focus $S\left(\dfrac{u^2\cos\alpha\sin\alpha}{g}, -\dfrac{u^2\cos\alpha}{2g}\right)$, $y = \dfrac{u^2}{2g}$ (equation of directrix).

4. Time of flight $= T = \dfrac{2u\sin\alpha}{g}$, $R = \dfrac{u^2}{g}\sin 2\alpha$, $R_{\max} = \dfrac{u^2}{g}$, when $\alpha = \dfrac{\pi}{4}$, $h = \dfrac{u^2\sin^2\alpha}{2g}$

5. Velocity at any point $v = \sqrt{(u\cos\alpha)^2 + (u\sin\alpha - gt)^2}$

6. Projected up the inclined plane $R = \dfrac{2u^2\cos\alpha\sin(\alpha-\beta)}{g\cos^2\beta}$,

$R_{\max} = \dfrac{u^2}{g(1+\sin\beta)}$, $T = 2u\sin(\alpha+\beta)/g\cos\beta$.

7. Projected down the inclined plane $R = \dfrac{2u^2\cos\alpha\sin(\alpha+\beta)}{g\cos^2\beta}$,

$R_{\max} = \dfrac{u^2}{g(1-\sin\beta)}$, $T = 2u\sin(\alpha+\beta)/g\cos\beta$.

Bibliography

Bellman, R and Dreyfus, *Applied Dynamics Programming* (Princeton Univ. Press).

Gant, P, *Mechanics IV* (Oxford Univ. Press).

Hassel blatt, B and Katok, A, *First Course in Dynamics* (Cambridge Univ. Press).

Hilderbrand, FB, *Method of Applied Mathematics* (Prentice Hall NY).

Irving and Milleux, *Mathematics in Physics and Engg.* (Academy Press NY).

Raisinghania, *Text Book on Dynamics* (Schand & Co.).

Sharma RD, *Mathematics II* (Dhanpat Rai Pvt Ltd.).

Singh Kuldip, *Monogram on Central Orbit* (NDA Publication).

Stephen lee, *An Introduction to Mathematics to Engg.* (Mechanics) (Hodder Publication UK).

White DJ, *Dynamics Programming*, (Addison Wesley, NY).

Index